Raymond J Puckholt

Computer Techniques
in Image Processing

Computer Techniques in Image Processing

HARRY C. ANDREWS

Department of Electrical Engineering
University of Southern California
Los Angeles, California

with
Contributions by

William K. Pratt

Department of Electrical Engineering
University of Southern California
Los Angeles, California

and

Kenneth Caspari

ITT Electrophysics Laboratories
Hyattsville, Maryland

ACADEMIC PRESS   New York   San Francisco   London   1970

A Subsidiary of Harcourt Brace Jovanovich, Publishers

COPYRIGHT © 1970, BY ACADEMIC PRESS, INC.
ALL RIGHTS RESERVED.
NO PART OF THIS PUBLICATION MAY BE REPRODUCED OR
TRANSMITTED IN ANY FORM OR BY ANY MEANS, ELECTRONIC
OR MECHANICAL, INCLUDING PHOTOCOPY, RECORDING, OR ANY
INFORMATION STORAGE AND RETRIEVAL SYSTEM, WITHOUT
PERMISSION IN WRITING FROM THE PUBLISHER.

ACADEMIC PRESS, INC.
111 Fifth Avenue, New York, New York 10003

United Kingdom Edition published by
ACADEMIC PRESS, INC. (LONDON) LTD.
24/28 Oval Road, London NW1

LIBRARY OF CONGRESS CATALOG CARD NUMBER: 74-117104

PRINTED IN THE UNITED STATES OF AMERICA

Contents

Preface	vii
Acknowledgments	ix

1. INTRODUCTION — 1

2. OPTICAL DATA PROCESSING

2.0 Introduction	5
2.1 Fourier Optics	6
2.2 Image Enhancement	19
2.3 Conclusions	28
References	28

3. DIGITAL OPTICAL PROCESSING

3.0 Implementation	31
3.1 Image Enhancement	36
3.2 Computer Calculated Diffraction Patterns	40
3.3 Digital Fourier Holography	50
3.4 Conclusions	53
References	54

4. TWO-DIMENSIONAL MATCHED FILTERING

4.0 Introduction	55
4.1 Theory	56
4.2 Implementation	59
4.3 Applications	64
4.4 Conclusions	69
References	70

5. ORTHOGONAL TRANSFORMATIONS
by Harry Andrews and Kenneth Caspari

5.0 Introduction	73
5.1 Fast Transformations	76
5.2 Generalized Spectral Analysis	90
5.3 Pattern Recognition Applications	98
5.4 Conclusions	101
References	102

6. IMAGE TRANSFORMS
by Harry Andrews and William Pratt

6.0 Introduction	105
6.1 Image Transformations	113
6.2 Analysis of Image Transforms	126
6.3 Conclusions	131
References	131

7. IMAGE CODING
by William Pratt and Harry Andrews

7.0 Introduction	135
7.1 Quantization of Image Transforms	136
7.2 Bandwidth Reduction	151
7.3 Error Tolerance	170
7.4 Conclusions	178
References	179
Author Index	181
Subject Index	184

Preface

This book is concerned with image processing utilizing the digital computer. The author hopes that this introduction to the subject will allow the spectrum of scientists dealing with data in two-dimensional formats (astronomers, geologists, biomedical engineers, etc.) to avail themselves of the present technology and perhaps in turn to further contribute to the field.

To approach this book the reader should be familar to some degree with conventional linear systems theory. However, background material is presented on Fourier optics and optical data processing for both coherent and incoherent sources.

The book is the outgrowth of a one-week summer short course, "Recognition of Two-Dimensional Images and Image Processing," offered at the University of Southern California in 1969. The content is developed in the venacular of electrical engineering with attempts made to coordinate with the language of related disciplines. Much of the optical material used as background originates from a one-semester graduate course on optical data processing which the author taught in the spring of 1969. All of the image processing results presented here have been implemented in the Electronic Images Processing Laboratory of the Department of Electrical Engineering at USC.

Acknowledgments

The author would like to thank Dr. Zohrab Kaprielian for his encouragement and enthusiasm for work on this project. In addition, the interest shown by Drs. Lloyd Welch, Julius Kane, and Irving Reed of the Electrical Engineering faculty is greatly appreciated. Finally, the guidance of Dr. William Pratt in the initial phases of the research presented in this book is enthusiastically acknowledged as are his contributions to Chapters 6 and 7.

Acknowledgments and special thanks go to Mr. Fred Billingsley and Mr. Thomas Rindfleisch of the Jet Propulsion Laboratory who so generously supplied the original digitized photographs of the moon scenes. In addition, their generous contractual support for research in the area of image coding for spacecraft applications was a major impetus to the completion of some of the work presented here.

The chapter on generalized spectral analysis and orthogonal transformations was motivated in part by the interest and encouragement of Mr. Joseph Lee and Mr. Ken Caspari of the ITT Electrophysics Laboratories in Hyattsville, Maryland. Much of the material and illustrations dealing with the generalized spectrum analyzer was provided by their computing laboratory and their support is certainly appreciated. In particular, the three-dimensional displays are especially useful.

The majority of the programming effort was provided by Mr. David Ketcham, an invaluable programmer as well as friend. The typing task was distributed throughout the Electronic Sciences Laboratory technical typing pool at USC and the precise and accurate results of their work is especially valued. Finally, a personal note of gratitude goes to my wife without whose encouragement and forbearance this volume would not have been written.

Computer Techniques
in Image Processing

Chapter 1 INTRODUCTION

A book entitled "Computer Techniques in Image Processing" by its very nature can cover a tremendously wide range and variety of diverse subjects which often could appear unrelated and independent. There is always the danger of covering such subjects and fields in too great a depth and in so doing losing the context and objective of such a book. This volume is heavily oriented toward the use of digital computers in image processing and visual pattern recognition techniques, and while the computer can be considered as only a tool in the process, it will provide a foundation and basis from which we can draw together the diverse fields which provide theoretical groundwork for much of the material presented here. The subject of image processing and visual pattern recognition is indeed a grandiose one, portions of which as yet remain more of an art rather than a science. Because each one of us is a visual pattern recognition expert in a human sense and each of us views images as subjectively as our varied backgrounds, it is not surprising to find the literature presenting as diverse a number of approaches as authors working in the field. As with any emerging "scientific" field, heuristic techniques play a major role in the development of sound theories from which future processing can be based. However, because of the diversity of individuals in the field and because of the lack of commonality of language and communication between various researchers, little if any theoretical ground can be made from correlating diverse heuristic approaches to specific pattern recognition problems. As an example,

2 INTRODUCTION

the medical expert classifying chromosomes applies techniques which initially appear totally unrelated to those used by the photo-reconnaissance interpreter searching for given military targets. In order that such classification procedures be placed on firmer ground, a theory must emerge which has applicability over the entire spectrum of two-dimensional pattern recognition and image processing problems. It is hoped that this book will provide the basis for the contribution to such a theory. Thus the philosophy of the book is not to present heuristic approaches which would often provide encouraging results for a given problem, but to present various theoretical backgrounds which could possibly contribute to an overall understanding or at least provide a common basis of approach from which heuristic techniques could be related.

The various fields from which this book arises include that of Fourier optics, two-dimensional matched filtering, the concepts of generalized spectral analysis, the theory of image coding for bandwidth reduction and noise immunity, and stochastic modeling techniques. The theory of Fourier optics and optical data processing for both coherent and incoherent illumination is presented as a foundation and background for conventional imaging systems. The presentation is based upon the assumption that the reader has a basic familiarity with conventional linear systems theory. Optic techniques for image enhancement are described with particular emphasis on earlier results of diffraction limited imaging systems.

The underlying concept of ultimate computer implementation prevails in the presentation. Consequently, additional descriptive material is provided in the areas that are particularly pregnant for computer simulation and solution, one example being the topic of superresolution. In addition, specific examples of digital computer simulations covering far field antenna diffraction patterns, digital holography, and image enhancement are included.

The study of two-dimensional matched filters for image processing is introduced in its historical application in radar and communication systems. The conventional one-dimensional matched filter results are quickly generalized to two-dimensions and a section on implementation follows. Three different implementation techniques are presented, one of which is entirely optical, one uses a combined optical and digital technique, and the third is entirely digital. The digital technique provides the greatest signal-to-noise ratio but requires a corresponding greater length of time for implementation. The application areas of two-dimensional matched filters are discussed with emphasis given to detection in a pattern recognition environment, gradient filters for emphasis of edge information, and a possible tool for image evaluation.

The material next presented is less oriented toward the optical image analogy but is directed toward the discovery of new horizons and techniques for two-dimensional processing in a completely digital environment. One such application is the emerging study of generalized spectral analysis. The material includes investigation of the optimum eigenvector decomposition for bandwidth reduction and leads into a discussion of efficient algorithmic implementation of a class of orthogonal transformations. This discussion includes the techniques for the fast Hadamard transform and fast Fourier transform as subsets of a much larger class of transformations obtained from kronecker products of matrices. A brief description of some of the applications of orthogonal decompositions to pattern recognition problems is presented with emphasis on the theories of maximizing variances and principle component analysis.

The theory of image coding is developed specifically for the Fourier transform of images in analogy with optical systems. The theory is generalized to any orthogonal transform and applications include specific transforms from those mentioned in the preceding section. The Fourier and Hadamard or Walsh transforms are particularly adapted to image processing and they are investigated for the quantization requirements necessary for the image transform domains. Bandwidth reduction techniques using the transform domains are suggested and experimentally verified in both threshold and zonal filtering cases. Finally, the application of error correcting techniques for improved noise immunity are described with specific examples for a binary symmetric channel error source.

It is hoped that the philosophy presented and the material to follow provide a reasonable approach and background to a subject that is becoming increasingly important with the increased power of digital computers. Because of the limitation of space, not all subjects and theories will be presented in proper depth and various oversights may occur. However, by the end of the book, the reader should have grasped some techniques and theoretical foundations from which he can develop his own image processing methods for specific systems. The subject has purposely been presented in such a way that any interested reader should have little difficulty in understanding most of the material. However, because of the diversity of theories utilized in various portions of the book, it is expected that not all chapters will be as easily digested as others for all readers. Consequently, extensive biliographies are included at the end of each chapter as convenient references for those interested in elaboration or clarification of specific material or of a specialized topic.

Chapter 2 OPTICAL DATA PROCESSING

2.0 INTRODUCTION

The use of high speed digital computers in the implementation of two-dimensional image processing and pattern recognition techniques is a relatively new approach to the classical problem of image enhancement and image detection. Examples of such techniques run the spectrum of computer restoration of atmospherically degraded images to computer recognition of biomedical objects. As in the development of any new field, it is often instructive and profitable to borrow from more established disciplines in the search for superior techniques which might be applicable to this form of computer processing. Such fields of particular interest for cross-pollination include two-dimensional linear systems theory, Fourier optics theory, the theory of optical data processing, and communication and detection theory. Toward this end it will be educational to review some of the elements of these theories in order to provide a firm foundation upon which to build an understanding and synthesis ability for digital computer implementation of specific two-dimensional processing tasks.

2.1 FOURIER OPTICS

2.1.1 Two-Dimensional Linear Systems

In the recent past, electrical engineering techniques have been successfully applied to the analysis of optical systems in which linear systems theory becomes immediately applicable [1-4]. The systems approach to optics and optical data processing allows a two-dimensional Fourier analysis in the study of diffraction theory and imaging systems. A system which is assumed to be linear implies that the total response is equal to a combination of more elementary responses and consequently a superposition property holds. If $\mathscr{H}$ is the mathematical operator describing the linear system, then the superposition property implies that

$$\mathscr{H}\{af_1(x,y) + bf_2(x,y)\} = a\mathscr{H}\{f_1(x,y)\} + b\mathscr{H}\{f_2(x,y)\} \qquad (1)$$

where a and b are real or complex constants.

Because the system is linear, it becomes useful to decompose the input function into a sum of displaced or shifted Dirac delta functions[1] Consequently, if the input is given by $f(x,y)$, then

$$f(x,y) = \iint_{-\infty}^{\infty} f(\xi,\eta)\,\delta(x-\xi, y-\eta)\,d\xi\,d\eta \qquad (2)$$

where $f(\xi,\eta)$ is the numerical weighting factor of the delta function at $\xi = x, \eta = y$. The output of the system then becomes

$$\mathscr{H}\{f(x,y)\} = \mathscr{H}\left\{\iint_{-\infty}^{\infty} f(\xi,\eta)\,\delta(x-\xi, y-\eta)\,d\xi\,d\eta\right\} \qquad (3)$$

but due to linearity, the system operator can be passed through the integration yielding

$$\mathscr{H}\{f(x,y)\} = \iint_{-\infty}^{\infty} f(\xi,\eta)\,\mathscr{H}\{\delta(x-\xi, y-\eta)\}\,d\xi\,d\eta \qquad (4)$$

---

[1] The theory of generalized functions in which the delta function finds application and mathematical justification can be reviewed in Lighthill [5].

where $\mathscr{L}\{\delta(x-\xi, y-\eta)\}$ is normally referred to as the impulse response of the system or the point spread function of the system in optical engineering terminology:

$$h(x, y; \xi, \eta) = \mathscr{L}\{\delta(x-\xi, y-\eta)\} \tag{5}$$

The linear system is said to be shift invariant if a shift in the input results in an equal shift in the output. A linear imaging system is space invariant (isoplanatic) if its impulse response, point spread function, depends only on $x-\xi, y-\eta$. Consequently,

$$h(x, y; \xi, \eta) = h(x-\xi, y-\eta) \tag{6}$$

and the superposition integral of Eq. (4) becomes a convolution integral

$$g(x, y) = \mathscr{L}\{f(x, y)\} = \int\!\!\!\int_{-\infty}^{\infty} f(\xi, \eta)\, h(x-\xi, y-\eta)\, d\xi\, d\eta \tag{7a}$$

or

$$g(x, y) = f(x, y) \circledast h(x, y) \tag{7b}$$

where $\circledast$ implies the two-dimensional convolution operation. The convolution theorem [1, 2] in Fourier analysis allows a representation of Eqs. (7a) or (7b) in terms of Fourier transforms.

$$G(u, v) = F(u, v)\, H(u, v) \tag{8}$$

where a capital lettered function implies the Fourier transform of a lower case function.

$$H(u, v) = \int\!\!\!\int_{-\infty}^{\infty} h(x, y)\, \exp\{-j(ux+vy)\} \tag{9a}$$

or

$$H(u, v) = \mathscr{F}\{h(x, y)\} \tag{9b}$$

where $\mathscr{F}\{\cdot\}$ is the two-dimensional Fourier transform operator. In this case $H(u, v)$ will become the coherent transfer function of an optical imaging system and is known as the system response function to most engineers. The physical operation of convolution as described by Eq. (7) can be interpreted as the integral of the product of two functions, one of which has been rotated 180 degrees and shifted by

8 OPTICAL DATA PROCESSING

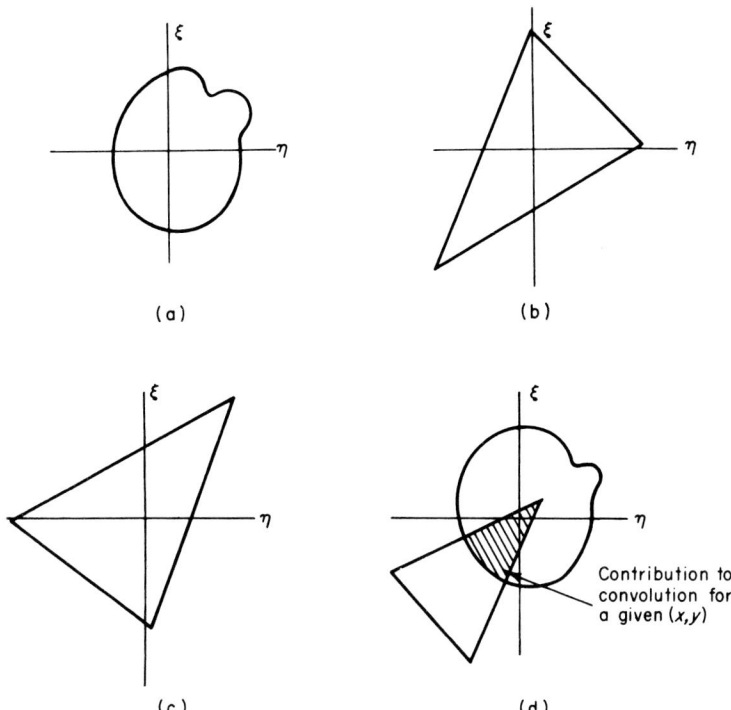

Fig. 1. Two-dimensional convolution. (a) $f(\xi,\eta)$; (b) $h(\xi,\eta)$; (c) $h(-\xi,-\eta)$; (d) $f(\xi,\eta)\,h(x-\xi, y-\eta)$.

an amount (x, y). Figure 1 offers a pictorial display of such an operation.

Because the convolution theorem allows the Fourier transform of the output of a linear shift invariant system to be equated to the Fourier transforms of the input multiplied by the system response, it becomes useful to mention a few properties of such a transformation. The transformation operation of Eq. (9a) decomposes a two-dimensional function into an infinite set of coefficients of two-dimensional orthogonal complex trigonometric waveforms given by the Fourier kernel $\exp\{-j(ux + vy)\}$. The transformation is symmetric conjugate (hermitian) and consequently the inverse Fourier transform is given by

$$h(x, y) = \iint_{-\infty}^{\infty} H(u, v) \exp\{j(ux + vy)\} \qquad (9c)$$

or

$$h(x, y) = \mathscr{F}^{-1}\{H(u, v)\} \qquad (9d)$$

to within a multiplicative constant.[2] The set of relationships given by Eqs. (9a)–(9d) describe $h(x, y)$ and $H(u, v)$ as Fourier transform pairs, $h(x, y) \leftrightarrow H(u, v)$. Other properties of the Fourier operation that will be useful are

Linearity:

$$\mathscr{F}\{af_1(x, y) + bf_2(x, y)\} = aF_1(u, v) + bF_2(u, v) \qquad (10a)$$

Scaling:

$$\mathscr{F}\{f(ax, by)\} = \frac{1}{|ab|} F\left(\frac{u}{a}, \frac{v}{b}\right) \qquad (10b)$$

Shift Property:

$$\mathscr{F}\{f(x - a, y - b)\} = F(u, v) \exp\{-j(ua + vb)\} \qquad (10c)$$

Convolution:

$$\mathscr{F}\{f(x, y) \circledast h(x, y)\} = F(u, v) H(u, v) \qquad (10d)$$

Autocorrelation:

$$\mathscr{F}\left\{\iint_{-\infty}^{\infty} f(\xi, \eta) f^*(\xi - x, \eta - y) \, d\xi \, d\eta\right\} = |F(u, v)|^2 \qquad (10e)$$

Parseval's Theorem:

$$\iint_{-\infty}^{\infty} f(x, y) g^*(x, y) \, dx \, dy = \iint_{-\infty}^{\infty} F(u, v) G^*(u, v) \, du \, dv \qquad (10f)$$

---

[2] This constant, either $1/4\pi^2$ or $1/2\pi$ depending on the definition of Eq. (9a), is ignored in this text for three reasons. It does not contribute to the basic understanding of linear systems, it does not exist in the physical world of optics, and it is easily ignored in the digital computer.

Gradients:
$$\mathscr{F}\{\nabla^2 f(x, y)\} = -(u^2 + v^2) F(u, v) \tag{10g}$$

Inversion:
$$\mathscr{F}\{\mathscr{F}^{-1}\{f(x, y)\}\} = \mathscr{F}^{-1}\{\mathscr{F}\{f(x, y)\}\} = f(x, y) \tag{10h}$$

Rotation:
$$\mathscr{F}\{\mathscr{F}\{f(x, y)\}\} = f(-x, -y) \tag{10i}$$

It is interesting to note that the Fourier transform is just one of an infinite number of ways of decomposing a signal into a set of coefficients of orthogonal waveforms. Any orthogonal transformation will provide a similar decomposition. The advantage of the Fourier technique is that the orthogonal waveforms (trigonometric functions) are the eigenvector solutions of linear systems and linear systems can often be used to describe natural processes. However when we are dealing with two-dimensional circularly symmetric systems, the eigenfunctions become Bessel functions and it becomes necessary to study Hankel transforms.

The motivation for the study of Hankel transforms is the natural circular symmetry of many optical systems. The Hankel transform is the two-dimensional Fourier transform of a circularly symmetric function and the transformed result is also circularly symmetric. A function is circularly symmetric if

$$f_1(x, y) = f((x^2 + y^2)^{1/2}) \tag{11}$$

By changing to a cylindrical coordinate system and assuming Eq. (11), the Hankel transform can be shown to be

$$F(\omega) = \int_0^\infty r f(r) J_0(r\omega) \, dr \tag{12a}$$

or

$$F(\omega) = \mathscr{H}\{f(r)\} \tag{12b}$$

where

$$r = (x^2 + y^2)^{1/2}$$

and

$$\omega = (u^2 + v^2)^{1/2}$$

The Hankel transform is equal to its own inverse and consequently

$$\mathcal{H}\{\mathcal{H}^{-1}\{f(r)\}\} = \mathcal{H}\{\mathcal{H}\{f(r)\}\} = f(r) \qquad (13)$$

Properties of the Hankel transform are similar to those of the Fourier transform a few of which appear below.

Scaling:

$$\mathcal{H}\{f(a\,r)\} = \frac{1}{a^2} F\left(\frac{\omega}{a}\right) \qquad (14a)$$

Gradient:

$$\mathcal{H}\{\nabla^2 f(r)\} = -\omega^2 F(\omega) \qquad (14b)$$

Convolution:

$$\mathcal{H}\{f(r) \circledast h(r)\} = F(\omega) H(\omega) \qquad (14c)$$

Parseval's Theorem:

$$\int_0^\infty r f(r)\, g^*(r)\, dr = \int_0^\infty \omega F(\omega)\, G^*(\omega)\, d\omega \qquad (14d)$$

The eigenfunctions of circular systems can now be shown to be zero-order Bessel functions of the first kind, $J_0(ar)$ [6]. This is due to the orthogonality property of such Bessel functions. Let the input function to the system be given by

$$f(r) = J_0(ar) \qquad (15a)$$

Then

$$F(\omega) = \frac{\delta(\omega - a)}{a} \qquad (15b)$$

and

$$G(\omega) = H(\omega) \frac{\delta(\omega - a)}{a} = \frac{H(a)}{a} \delta(\omega - a) \qquad (15c)$$

Consequently

$$g(r) = \frac{H(a)}{a} J_0(ar) \qquad (15d)$$

Therefore when $J_0(ar)$ is input to a circularly symmetric linear sytem, it also is the output of such a system modified by a constant eigenvalue given by $H(a)/a$. Because the system transform and its inverse are both identical, $J_0(ar)$ must be an eigenvector of that system.

2.1.2 Diffraction Theory

With the above brief introduction to two-dimensional linear systems theory, it is now appropriate to investigate some of the elements of diffraction theory in the search for an understanding of some of the processes and limitations of physical optical systems to theoretical imaging and data processing at optical frequencies. The definition of diffraction for purposes here will be the deviation of light rays from geometrically predicted paths and only the scalar theory of light will be applicable. This assumption will imply that all theories developed will be based on the scalar amplitude of one transverse component of either the electric or magnetic field of light. Consequently, the coupling of the two fields by Maxwell's equations will be neglected. The scalar theory is valid for diffracting apertures that are large with respect to the optical wavelength; "diffraction limited" systems will mean no other limitations than those offered by diffraction theory, which implies the system is perfect and only natural phenomena, uncontrollable by man, is a limitation. While such a physical limitation results from the theory of diffraction, there are certain mathematical techniques recenty developed which imply resolution is possible beyond the classical limit. Such techniques are especially exciting when envisioned in the context of digital computer image processing and will be discussed in the section on image enhancement.

It is the objective of this section not to develop rigorous diffraction theory but to present the Fresnel and Fraunhofer diffraction formulas in a form quite similar to two-dimensional Fourier transforms. The formulas will be based upon the Huygens' Fresnel principle given by the Rayleigh–Sommerfeld model for coherent illumination [7], where $k = 2\pi/\lambda$ and λ is the wavelength of light

$$U(P_i) = \frac{1}{j\lambda} \iint_s \frac{U(P_a)}{r_{ia}} \exp\{jkr_{ia}\} \cos(\bar{n}\bar{r}_{ia}) \, ds \qquad (16)$$

where the geometry for this equation is given in Fig. 2. The equation

2.1 FOURIER OPTICS

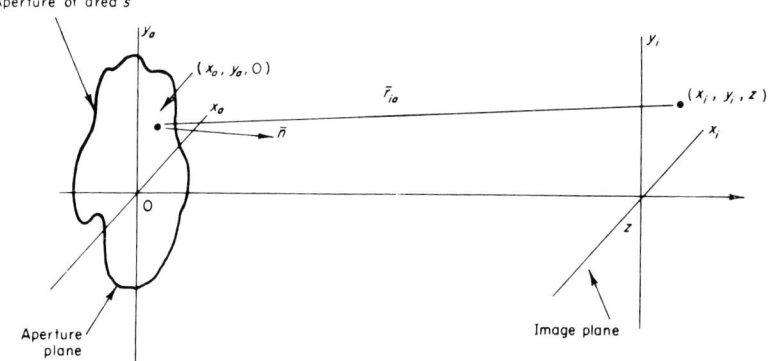

Fig. 2. Rayleigh–Sommerfeld diffraction model.

states that the field at point $P_i = (x_i, y_i, z)$ in the image plane is equal to $1/j\lambda$ times the integral of the field $U(P_a)$ at any point in the aperture $P_a = (x_a, y_a, 0)$ multiplied by a phase factor and angular variation which are both functions of the distance from the point of observation, P_i, and the point of integration, P_a, as that point explores the entire aperture. By assuming that the image plane is located a distance z much larger than the dimensions of the aperture and the dimensions of interest in the image plane, the following approximations can be made.

$$\cos(\bar{n}, \bar{r}_{ia}) = 1$$
$$r_{ia} = z$$

The distance between points of integration and observation, r_{ia}, can be expressed in terms of P_a, and P_i as

$$r_{ia} = (z^2 + (x_i - x_a)^2 + (y_i - y_a)^2)^{1/2}$$

and

$$r_{ia} = z\left(1 + \left[\frac{(x_i - x_a)}{z}\right]^2 + \left[\frac{(y_i - y_a)}{z}\right]\right)^{1/2}$$

and by retaining only the first two terms of the binomial expansion

14 OPTICAL DATA PROCESSING

of the square root, r_{ia}, can be approximated by

$$r_{ia} = z\left[1 + \frac{(x_i - x_a)^2}{2z^2} + \frac{(y_i - y_a)^2}{2z^2}\right]$$

Using the above assumptions and approximations the Rayleigh-Sommerfeld model results in the following Fresnel diffraction equation

$$U(x_i, y_i, z) = \frac{\exp\{jkz\}}{j\lambda z}$$
$$\times \iint_s U(x_a, y_a, 0) \exp\left\{j\frac{k}{2z}[(x_i - x_a)^2 + (y_i - y_a)^2]\right\} ds$$
(17a)

It is obvious that this equation is a finite convolution. The field in the image plane is given by the multiplicative constant, $\exp\{jkz\}/j\lambda z$ (a function of the placement of the image plane, z, and the wavelength of light, λ) times the convolution of the field in the aperture, $U(x_a, y_a)$, with the phase factor

$$\exp\left\{\frac{jk}{2z}(x_a^2 + y_a^2)\right\}$$

The Fresnel diffraction equation (17a) can be expanded and represented as

$$U(x_i, y_i, z) = \frac{\exp\{jkz\}}{j\lambda z} \exp\left\{j\frac{k}{2z}(x_i^2 + y_i^2)\right\}$$
$$\times \mathscr{F}\left\{U(x_a, y_a, 0) \exp\left\{j\frac{k}{2z}(x_a^2 + y_a^2)\right\}\right\} \quad (17b)$$

where $\mathscr{F}\{\cdot\}$ is the two-dimensional Fourier transform operator described earlier.

To obtain the Fraunhofer diffraction formula from the Fresnel Eqs. (17a) or (17b) a further assumption will be made. If the image plane is placed so that

$$z \gg \frac{k}{2}(x_a^2 + y_a^2)$$

then the field in that plane becomes

$$U(x_i, y_i, z) = \frac{\exp\{jkz\}}{j\lambda z} \exp\left\{j\frac{k}{2z}(x_i^2 + y_i^2)\right\} \mathscr{F}\{U(x_a, y_a, 0)\}$$

(18)

which is the Fraunhofer diffraction formula desired. Note that for intensity-sensitive optical instruments, the intensity diffraction pattern is proportional to the magnitude of the Fourier transform of the field at the aperture. With this brief description of the Fresnel or near field and Fraunhofer or far field diffraction patterns, it is now appropriate to apply the results to both coherent and incoherent imaging systems.

2.1.3 Imaging Systems

It is instructive to point out at this time that an underlying assumption which has been prevalant in the derivations for diffraction theory is the concept of linearity. For both the Fresnel and Fraunhofer results, the image plane waveform is a linear combination, convolution

TABLE I

Light wave $U(x, y)$	Linear system application
Passing through	
1. a plane, $f(x, y)$	$g(x, y) = U(x, y) f(x, y)$
2. a cylindrical lens	$g(x, y) = U(x, y) \exp\left\{-j\frac{k}{2f} x^2\right\}$
	or
	$g(x, y) = U(x, y) \exp\left\{-j\frac{k}{2f} y^2\right\}$
3. a convex spherical lens	$g(x, y) = U(x, y) \exp\left\{-j\frac{k}{2f}(x^2 + y^2)\right\}$
4. a conical lens in the x-direction	$g(x, y) = U(x, y) \exp\left\{-j\frac{kx^2}{h(y)}\right\}$
5. a change of axial position d	$g(x, y) = U(x, y) \circledast \exp\left\{\frac{k}{2d}(x^2 + y^2)\right\}$

16 OPTICAL DATA PROCESSING

or Fourier integral respectively, of the aperture plane waveform. Consequently diffraction theory under the assumption of coherent illumination is linear in phase and amplitude. Thus, it is not unreasonable to expect that coherent imaging systems will also be linear in phase and amplitude. Using this concept it becomes expedient to describe coherent imaging systems in terms of one and two-dimensional linear functional operations on an input. Such an approach is described by Vander Lugt [8] and is briefly reviewed here. Table I indicates a few optical elements of interest.

Entry 5 from the table shows that a change in axial position causes a convolution of the input wave with a quadratic phase factor which

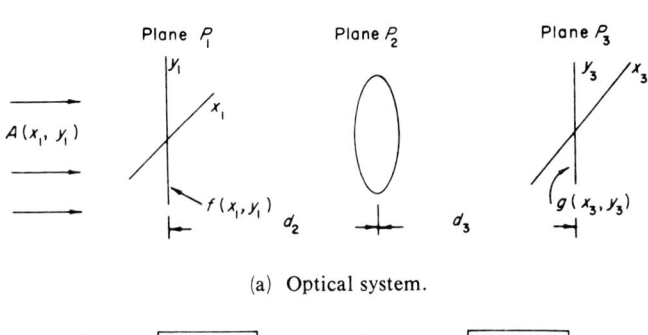

Fig. 3. A possible imaging system.

is a function of the distance changed. However this is the Fresnel diffraction result derived earlier in Eq. (17a). Entry 3 indicates that a convex or positive spherical lens is equivalent to multiplication of the input waveform by a quadratic phase factor. Noting that the sign of the exponential of the quadratic phase factor due to the lens is opposite that due to the change in axial position, it should be possible to have the two terms cancel. This in fact does result in the imaging situation described by the optical and linear system counterpart of Fig. 3. In this figure both the optical and equivalent linear systems are presented as well as the mathematical expression for the output.

$$g(x_3, y_3) = \left[\{[A(x_1, y_1)f(x_1, y_1)] \circledast h_2(x_1, y_1)\} \times \exp\left\{-j\frac{k}{2f}(x_2^2 + y_2^2)\right\} \right] \circledast h_3(x_2, y_2) \quad (19a)$$

If the input wavefunction is assumed to be a uniform, unit-amplitude plane wave, then $A(x_1, y_1) = 1$. If $d_2 + d_3 = f$ in Fig. 3, then the output function becomes

$$g(x_3, y_3) = \exp\left\{\frac{jk}{2}\left(d_3 + \frac{d_3^2}{d_2}\right)(x_3^2 + y_3^2)\right\} f\left(-\frac{d_3 x_3}{d_2}, -\frac{d_3 y_3}{d_2}\right) \quad (19b)$$

The mathematical details are left to the reader or can be obtained from Vander Lugt [8]. Aside from the quadratic phase constant, the output is seen to be equal to the function $f(x, y)$ rotated 180° and scaled by d_3/d_2. This is the imaging condition for the optical system of Fig. 3.

If the input waveform is still assumed to be a uniform plane wave but now both d_3 and d_2 are set equal to the focal length of the lens, f, then the output becomes

$$g(x_3, y_3) = \iint_{-\infty}^{\infty} f(x_1, y_1) \exp\left\{-j\frac{k}{f}(x_1 x_3 + y_1 y_3)\right\} dx_1 \, dy_1 \quad (19c)$$

which is the Fourier transform condition for the optical system of Fig. 3. Again the details are derivable using the concepts of two-dimensional convolutions and Fourier analysis.

18 OPTICAL DATA PROCESSING

To this point we have only considered a particular optical system, Fig. 3, and it now becomes necessary to consider generalized optical systems. Such systems will be characterized by their input/output or terminal properties known as entrance or exit pupils. Using the concepts of linearity, it can be shown that the impulse or point spread function of a coherently illuminated optical system is equivalent to the Fourier transform of the smallest aperture of the system projected onto the exit pupil [9]

$$h(x, y) = \mathscr{F}\{P(u, v)\} \quad (20)$$

where $P(u, v)$ is a 0-1 real function having the value 1 in the area of the aperture and 0 elsewhere for a diffraction limited system. The system response or coherent transfer function of the system is then given by

$$H(u, v) = \mathscr{F}\{\mathscr{F}\{P(u, v)\}\} = P(-u, -v) \quad (21)$$

and will be abbreviated as CTF and the rotated coordinate system will hereafter be ignored. This statement then can be interpreted as relating the highest spatial frequency that the system can pass with the cutoff point or finite nature of the limiting aperture, $P(u, v)$, in the generalized coherent imaging system.

Until now the discussion has been entirely relevant to the context of coherent imaging systems. However, normal photographic processes are carried out under incoherent illumination conditions and our theory of two-dimensional image processing must be enlarged to encompass such systems. Fortunately, it can be shown that for totally incoherent illumination, diffraction limited optical systems are no longer linear in phase and amplitude, but become linear in intensity. However, the impulse response of such a system becomes the magnitude squared of the impulse response of the coherently illuminated system and the linearity relationship is expressed as

$$I_i(x_i, y_i) = \iint_{-\infty}^{\infty} |h(x_i - x_o, y_i - y_o)|^2 I_o(x_o, y_o) \, dx_o \, dy_o \quad (22)$$

where the subscripts (i, o) refer to "image" and "object" respectively. The system response of the incoherent system, known as the optical transfer function, OTF, when normalized becomes

$$\mathcal{H}(u, v) = \frac{\mathcal{F}\{|h(x, y)|^2\}}{\iint\limits_{-\infty}^{\infty} |h(x, y)|^2 \, dx \, dy} \tag{23a}$$

The optical transfer function, OTF, can be expressed in terms of the coherent transfer function, CTF, as

$$\mathcal{H}(u, v) = \frac{\iint\limits_{-\infty}^{\infty} H(\xi - u/2, \eta - v/2) \, H^*(\xi + u/2, \eta + v/2) \, d\xi \, d\eta}{\iint\limits_{-\infty}^{\infty} |H(\xi, \eta)|^2 \, d\xi \, d\eta}$$

(23b)

Thus the OTF is equivalent to the normalized autocorrelation of the CTF. This indicates that the OTF will always have twice the cutoff frequency of the CTF, fact which initially might appear quite appealing. However, the comparison is between an intensity system and a phase and amplitude system, and the situation can arise where one system is superior for one type of signal but inferior for another. Some immediate consequences due to the definition of the OTF are

(a) $\mathcal{H}(0, 0) = 1$
(b) $|\mathcal{H}(u, v)| \leq \mathcal{H}(0, 0)$
(c) $\mathcal{H}(-u, -v) = \mathcal{H}^*(u, v)$

and for diffraction limited, that is, aberration-free systems,

(d) $\mathcal{H}(u, v) = \mathcal{H}^*(u, v)$
(e) $\mathcal{H}(u, v) \geq 0$

In the next section, we shall see that certain aberrations cause the OTF to become negative for certain spatial frequency regions.

2.2 IMAGE ENHANCEMENT

The discussion thus far has been relevant only to analysis procedures. However, with this background it is hoped that certain synthesis techniques will come to mind. Probably the most obvious area for synthesis procedure is that of image enhancement. It has often been argued that if one knew the OTF of an imaging system, then a compensating filter could be obtained to reconstruct degradations due to the imaging process. While in theory such techniques are possible, in fact, until the introduction of digital computer techniques, such attempts have had limited success mainly due to noise and optical bench requirements as well as dynamic range and film storage problems.

2.2.1 Inverse Filters

The most obvious means of constructing a compensating filter is by multiplying the cumulative transfer function of an imaging system by its inverse function so that the resultant filtering effect is that of a unity gain zero phase shift filter. However, for the ideal diffraction limited coherent imaging system it was shown that the CTF was equal to the pupil function, Eq. (21), which is already a unity filter function in its passband. Consequently, for diffraction-limited coherent imaging systems, inverse filtering techniques are not useful and a super resolution technique, to be described later, must be employed for increased resolution. However, most imaging systems are far from diffraction limited and in fact have degradations due to a variety of causes, some of which might be aberrations, image blur due to motion, poor contrast, or atmosphere turbulence. By assuming that the image forming system is linear and stationary, an inverse filter can be approximated and considerable improvement obtained. Investigators [10] have reported that turbulence effects averaged over time tend to reduce contrast and if time-averaging (exposure) is long enough then the OTF of the imaging system tends to be Gaussian. Consequently, given a predetermined set of Gaussian OTF's the best inverse Gaussian filter can be selected as that which gives the maximum contrast in the filtered images. One technique that is particularly appealing and which some individuals have used [10, 11] is to record the OTF at the same time the image is formed. This can be done by recording the point spread function of a point source of light passing through the same degradations as the object information. Of course, this technique might not always be feasible in which case a computer search of the best inverse Gaussian filter could be resorted to. Such an approach has been made in the experimental results presented in Chapter 3.

One particular type of degradating phenomena inherent in an imaging system independent of any external atmospheric turbulence considerations is due to aberrations. While there are a variety of different types of aberrations, the simplest to handle mathematically are those which can be assumed to introduce a phase shift in the pupil function of a diffraction limited system. Consequently the CTF becomes

$$H(u, v) = P(u, v) \exp\{jk\, W(u, v)\} \tag{24a}$$

and the OTF can be represented as

$$\mathcal{H}(u, v) = \left[\int_{-\infty}^{\infty} P(\xi - u/2, \eta - v/2)\, P(\xi + u/2, \eta + v/2) \right.$$
$$\times \exp\{jk[W(\xi - u/2, \eta - v/2)$$
$$\left. - W(\xi + u/2, \eta + v/2)]\}\, d\xi\, d\eta \right] \Big/ \iint_{-\infty}^{\infty} (P(\xi, \eta))\, d\xi\, d\eta \tag{24b}$$

For the simple case of a defocused imaging system the phase error becomes

$$W(u, v) = \frac{\varepsilon(u^2 + v^2)}{2} \tag{25}$$

For a square aperture the OTF of Eq. (24b) can be found [1, 12] explicitly to be

$$\mathcal{H}(u, v) = \wedge(u) \wedge (v)\, \text{sinc}[Ku \wedge (u)]\, \text{sinc}[Kv \wedge (v)] \tag{26a}$$

to within a scale factor and where sinc[·] is the sin x/x function, K is a constant proportional to the amount of defocus, ε, and $\wedge(\cdot)$ is the triangle function

$$\wedge(x) = \begin{cases} 1 - |x|, & |x| \leq 1 \\ 0, & \text{otherwise} \end{cases}$$

22 OPTICAL DATA PROCESSING

For badly focused systems, the OTF approaches the two-dimensional sinc function.

$$\mathcal{H}(u, v) = \text{sinc}(Ku)\,\text{sinc}(Kv) \qquad (26b)$$

which is the prediction obtained by geometrical optic considerations. For a circular aperture and badly out-of-focused systems, geometrical optics predicts that

$$\mathcal{H}(\omega) = \frac{J_1(\varepsilon\omega)}{\varepsilon\omega} \qquad (26c)$$

where $J_1(\cdot)$ is the first-order Bessel function. It is interesting to note that the three OTF's described in Eq. (26) can become negative in certain spatial frequency bands cosequently introducing a 180° phase shift for certain frequencies. Also, zeros occur in the OTF's thus making it impossible to obtain finite reciprocal functions for compensating filters. However, a computer can be easily programmed to approximate the reciprocal functions with finite values.

One area in which inverse filtering has had considerable success is that of compensating for image blur due to some type of lateral motion [13–15]. Considering the case of one-dimensional motion, the image recorded on film can be shown to be a convolution of the undistorted image with a rectangular function whose width is proportional to the exposure time. Consequently, the effective OTF is the Fourier transform of the rectangle function which is a sinc($\cdot$) function.

$$\mathcal{H}(u, v) = \frac{\sin(au)}{u} \qquad (27a)$$

However, the inverse filter of this function also does not remain finite. Lohmann has suggested a shutter modulation technique to avoid the zeros in the optical transfer function [16]. However, one technique to compensate for the OTF of Eq. (27a) is to use a "differentiator" filter by multiplying by the ramp, u. The resulting transfer function becomes

$$\mathcal{H}(u, v) = \sin au \qquad (27b)$$

2.2 IMAGE ENHANCEMENT

The resulting image then becomes the convolution of the unblurred image with the Fourier transform of the sin function.

$$g(x, y) = f(x - a, y) - f(x + a, y) \qquad (27c)$$

to within a phase shift. Thus the unblurred image is recovered but with a "ghost" spaced a distance $x = 2a$ away. For large motion or small images the displacement will be large enough to ensure that the images will not overlap.

The compensating filter used for one-dimensional image blur was a ramp in the frequency domain which by Fourier analysis is equivalent to a differentiator in the x-dimension of the space domain. Higher order polynomials in the frequency domain have differential counterparts in the space domain. For instance the Laplacian operator, often used for edge detection and edge enhancement, is given by

$$\nabla^2 = \frac{\partial^2}{\partial x^2} + \frac{\partial^2}{\partial y^2}$$

which is equivalent to multiplying the frequency domain by $u^2 + v^2$. However, this is also equivalent to compensating for a circularly symmetric optical system whose response falls off as the square of the spatial frequency. Gradients, Laplacians, and higher order operators will be discussed in greater detail in the section on matched filters. However, they can be used as the compensating filters presented here.

Finally, it should be pointed out that some degrading effects in image construction are more realistically modeled as multiplicative effects rather than additive. In such cases, certain nonlinear filtering techniques [17] have been successfully applied. Specifically, a logarithmic preprocessor separates the multiplicative functions and allows conventional linear filtering theory to be employed. Then the output of the linear system is raised to an exponential power to restore the effect of the logarithmic operation. The success of this approach is due in part to the use of digital computers to perform the nonlinear mathematical operations exactly on the pre and post processed images.

2.2.2. Contrast Filters

A class of filters known as contrast filters are used to perform a contrast improvement in imaging systems. For most optical systems, the OTF is a passive process in the sense that amplification never occurs. However, it is possible to construct filters which provide an absolute increase in gain for a specific band of frequencies [18]. An image with low contrast can be approximated by a modulation function riding on a dc backgroud of light

$$f(x, y) = A + g(x, y) \tag{28}$$

For such a situation, an obvious technique for contrast improvement is an optical stop dc frequency, placed at the origin of the transfer function of the optical system.

A more sophisticated class of filters have been used to implement a phase contrast to obtain an intensity mapping of a pure phase image, $f(x, y) = \exp\{j\phi(x, y)\}$. The technique, known as Zernike's phase contrast method [19], is to assume a small phase variation in the image, $f(x, y)$, and approximate it by the first two terms of a power series expansion

$$f(x, y) = 1 + j\phi(x, y) \tag{29a}$$

By introducing a phase changing plate at the origin in the frequency domain, the dc term will become phase shifted allowing an intensity output to be proportional to $\phi(x, y)$. If the phase plate is made equal to $\pi/2$ radians, then the intensity image output becomes

$$g(x, y) = |\exp\{j\pi/2\} + j\phi(x, y)|^2 \tag{29b}$$

$$= 1 + 2\phi(x, y) \tag{29c}$$

where higher order terms of ϕ have again been set equal to zero.

The Schlieren method for observing phase objects is obtained by introducing a knife edge in the spatial frequency domain of an image [20-22]. However, a phase grating technique has also been used to observe phase objects with some success [23]. The knife edge technique is based on the concept of Hilbert transforms and can be analyzed in terms of quadrature filters if desired.

2.2.3 Superresolution

For many years, investigators working with optical systems assumed that the classical Rayleigh diffraction limit of resolution was a theoretical limit beyond which no higher spatial frequencies could be resolved. However, pioneering work in super-gain antennas and theoretical applications to optics (di Francia [24]) has led the way out of the myth that the diffraction limit is a theoretical absolute. A second paper by di Francia [25] applied an information theoretic approach to the problems of resolution in optical systems and showed that for super-resolution to be possible, some *a priori* knowledge concerning the object before it enters the imaging system must be known. It turns out that the *a priori* knowledge necessary is a finite limit in space of the original object. However, it is still a fact that diffraction limited systems act as low pass filters and for coherent optics, the cutoff frequency is given by the pupil function, Eq. (21), whereas for incoherent systems, the cutoff frequency is given by the autocorrelation of the pupil function, Eq. (23b). In either case, only a finite range of spatial frequencies are available. The question of increasing the resolution of the object by increasing the range of the spatial frequencies must then be answered by the theory of analytic continuation. Briefly, this theory states in optical terms that a finite range object has an infinite range spectrum which can be obtained from a finite range spectrum by analytic continuation of that spectrum and that the continuation is unique. Theoretically, then, an optical system with a finite pupil function can still attain as high a resolution as desired. Limitations of enormous current requirements in microwave antenna systems and large flux requirements in optical systems initially were quite discouraging. Another question which was initially unanswered was what technique of extrapolation of the frequency domain would be best in an implementation sense. The traditional techniques of polynomial plane fitting and approximation by rational functions tended to be ill-behaved at large distances from the known finite range spectra. An approach, due to sampling theory [26, 27], has proven successful where higher order sample points are calculated from the coefficients in the known spectral region. One-dimensional mathematical examples have indeed demonstrated that it is possible to resolve beyond the classical limit.

One approach to the extrapolation technique that has considerable mathematical appeal is through the use of prolate spheroidal wave functions [28]. Slepian and Pollak [29] and Landau and Pollak [30] have shown that the prolate spheroidal waveforms are complete on any closed interval and are thereafter defined over the entire real line. Specifically, it will be assumed that the one-dimensional Fourier transform, $F(u)$, of a finite object, $f(x)$ is known only on a finite interval $[-U/2, U/2]$ of the spatial frequency axis. Then $F(u)$ is a band-limited function, in the sense that it has a finite range limited transform, $f(x)$, and we can apply the results of Slepian and Pollak [29]. Specifically, the class of prolate spheroidal wave functions, $\varphi_i(u)$ are band-limited, orthonormal on the real line, and complete in any finite interval:

$$\int_{-\infty}^{\infty} \varphi_i(u)\varphi_j(u)\,du = \delta(i-j) \tag{30a}$$

$$\int_{-U/2}^{U/2} \varphi_i(u)\varphi_j(u)\,du = \lambda_i\,\delta(i-j) \tag{30b}$$

$$\lambda_i \varphi_i = \int_{-U/2}^{U/2} \frac{\sin \Omega(u-s)}{\pi(u-s)} \varphi_i(s)\,ds \tag{30c}$$

Note that Eq. (30c) indicates that the eigenvector solution to the convolution of a sinc function with a band-limited signal is the prolate spheroidal waveform. But for a one-dimensional coherently illuminated aperture, a rectangle is the equivalent pupil function. Consequently the one-dimensional image, $g(x)$, is the convolution of the transform of the rectangle, a sinc function, with the object, and the results of Eq. (30c) can be directly applied. The expansion of $F(u)$ can now be accomplished in terms of the prolate spheroidal wave functions.

$$F(u) = \sum_{n=0}^{\infty} a_n \varphi_n(u) \tag{31a}$$

where

$$a_n = \frac{1}{\lambda_n} \int_{-U/2}^{U/2} F(u)\varphi_n(u)\,du \tag{31b}$$

Consequently, the coefficients, $\{a_n\}$, necessary for complete definition of $F(u)$ on the entire line can be determined from the knowledge of $F(u)$ on the interval $[-U/2, U/2]$, Eq. (31b).

It is interesting to note that for computer implementation of the above process, a finite summation of Eq. (31a), must be used. The question of mean square truncation error over the infinite interval is [29]

$$\mathrm{mse}_\infty = \sum_{n=N+1}^{\infty} a_n^2 \qquad (32a)$$

and over the finite interval by

$$\mathrm{mse}_U = \sum_{n=N+1}^{\infty} a_n^2 \lambda_n \qquad (32b)$$

Because the eigenvalues are in decreasing order as a function of the index n [29], this mean square error is minimum in a Karhunen-Loeve expansion sense [31]. Finally, it should be mentioned that the question of violation of Heisenberg's uncertainty principle is covered by Landau and Pollak [30]. Various individuals have applied the concepts of using prolate spheroidal wave functions to one-dimensional imaging system with some success in their mathematical examples [32-34]. It has also been pointed out that very high side lobes are obtained in certain applications which must be avoided in practical systems [33]. In one paper of information theoretic interest, the results of the spheroidal waveform analysis to the definition of capacity and information rate for an imaging system have been applied [35]. Landau and Pollak [36] and Landgrebe and Cooper [37] have pointed out that the number of coefficients of Eq. (31a) for a given reconstruction accuracy are of the same order as that required using sampling theory extrapolation techniques [26, 27].

Finally, the question and partial answer to apodization can be explained in terms of the prolate spheroidal wave functions. Specifically, it is desired to find that filter, or lens coating in optical terms, to improve the point spread function, $h(x)$, to make it as narrow as possible. The question can be equivalently expressed as what function maximizes the ratio [30, 38].

$$R = \frac{\int_{-b}^{b} |h(x)|^2 \, dx}{\int_{-\infty}^{\infty} |h(x)|^2 \, dx} \qquad (33)$$

The resulting solution is given by the zeroth-order prolate spheroidal wavefunction, $\varphi_0(x)$, with the ratio of maximum energy enclosed in the bounded interval $[-b, b]$ given by λ_0.

2.3 CONCLUSIONS

This chapter has presented two of the basic tools, two-dimensional linear systems and Fourier optics, that we shall use for the generation of a computerized image processing system. Considerable effort has been devoted to the mathematics involved in the linear systems work with examples taken from the optical world where possible. Thus diffraction theory fits nicely into the context of linear systems; and although a rigorous derivation for the Rayleigh–Sommerfeld model, Eq. (16), for coherent illumination has not been developed, it is felt that the presentation of the Fresnel and Fraunhofer diffraction equation from such a starting point is well justified. The physical phenomena of Fresnel diffraction is shown to be a convolution operation in terms of linear systems theory. Equivalently, that of Fraunhofer diffraction is shown to be the Fourier transform operation, again in terms of linear systems theory. Imaging systems are then described using the concepts already put forth. The coherent and optical transfer functions (CTF, OTF) are defined in terms of pupil functions for both coherently and incoherently illuminated imaging systems, respectively. With the tools afforded by the theory of Fourier Optics, it was then possible to develop some image enhancement techniques, the first being the use of inverse filtering operations to compensate for optically degrading transfer functions in the original imaging system. Mathematical examples were given for out of focused images and motion blur. A brief review of contrast filters was present followed by the concepts of superresolution with possible applications for a computer environment. The following chapter illustrates a variety of applications for computerization of the principles set forth herein.

REFERENCES

1. J. W. Goodman, *Introduction to Fourier Optics*. McGraw-Hill, New York, 1968.

2. A. Papoulis, *Systems and Transforms with Applications in Optics.* McGraw-Hill, New York, 1968.
3. E. L. O'Neill, *Introduction to Statistical Optics.* Addison-Wesley, Reading, Massachusetts, 1963.
4. L. Mertz, *Transformations in Optics.* Wiley, New York, 1965.
5. M. J. Lighthill, *Introduction to Fourier Analysis and Generalized Functions.* Cambridge Univ. Press, London and New York, 1960.
6. A. Papoulis, *Systems and Transforms with Applications in Optics.* p. 152. McGraw-Hill, New York, 1968.
7. J. W. Goodman, *Introduction to Fourier Optics*, p. 42. McGraw-Hill, New York, 1968.
8. A. Vander Lugt, "Operational Notation for the Analysis and Synthesis of Optical Data-Processing System," *Proc. IEEE* **54**, No. 8, 1055-1063 (August 1966).
9. J. W. Goodman, *Introduction to Fourier Optics*, p. 105. McGraw-Hill, New York, 1968.
10. P. F. Mueller and G. O. Reynolds, "Image Restoration by Removal of Random-Media Degradation," *J. Opt. Soc. Am.* **57**, No. 11, 1338-1344 (November 1967).
11. B. L. McGlamery, "Restoration of Turbulence-Degraded Images," *J. Opt. Soc. Am.* **57**, No. 3, 293-297 (March 1967).
12. R. Barakat, "Application of the Sampling Theorem to Optical Diffraction Theory," *J. Opt. Soc. Am.* **54**, No. 7, 920-930 (July 1964).
13. A. Papoulis, *Systems and Transforms with Applications in Optics*, p. 436. McGraw-Hill, New York, 1968.
14. M. R. Schroeder, "Images from Computers," *IEEE Spectrum* 66-78 (March 1969).
15. J. L. Harris, "Image Evaluation and Restoration," *J. Opt. Soc. Am.* **56**, No. 5, 569-574 (May 1966).
16. A. W. Lohmann, *Recognition of Two-Dimensional Images and Image Processing.* July 28—August 1, 1969. In Lecture presented at USC short course.
17. A. V. Oppenheim, *et al.* "Nonlinear Filtering of Multiplied and Convolved Signals," *Proc. IEEE* **56**, No. 8, 1264-1291 (August 1968).
18. J. D. Armitage, A. W. Lohmann, and R. B. Herrick, "Absolute Contrast Enhancement," *Appl. Opt.* **4**, No. 4, 445-451 (April 1965).
19. J. W. Goodman, *Introduction to Fourier Optics*, p. 145. McGraw-Hill, New York, 1968.
20. A. Vander Lugt, "A Review of Optical Data-Processing Techniques," *Opt. Acta.* **15**, 1-33 (Jaunary/February 1968).
21. A. Papoulis, *Systems and Transforms with Applications in Optics*, p. 435. McGraw-Hill, New York, 1968.
22. J. W. Goodman, *Introduction to Fourier Optics*, p.193. McGraw-Hill, New York, 1968.
23. A. W. Lohmann and D. P. Paris, "Computer Generated Spatial Filters for Coherent Optical Data Processing," *Appl. Opt.* **7**, No. 4, 651-665 (April 1968).

24. G. T. di Francia, "Super-Gain Antennas and Optical Resolving Power," *Nuovo Cimento Suppl.* **9,** (9) 426 (1952).
25. G. T. di Francia, "Directivity, Super-Gain and Information," *IRE Trans. Antennas Propagation* **4,** No. 3, 473-478 (July 1956).
26. J. W. Goodman, *Introduction to Fourier Optics*, p. 133. McGraw-Hill, New York, 1968.
27. J. L. Harris, "Diffraction and Resolving Power," *J. Opt. Soc. Am.* **54,** No. 7, 931-936 (1964).
28. C. Flammer, *Spheroidal Wave Functions*. Stanford Univ. Press, Stanford, California, 1957.
29. D. Slepian and H. O. Pollak, "Prolate Spheroidal Wave Functions, Fourier Analysis and Uncertainty—I," *BSTJ* 43-63 (January 1961).
30. H. J., Landau and H. O. Pollak, "Prolate Spheroidal Wave Functions, Fourier Analysis and Uncertainty—II," *BSTJ*, 65-84 (January 1961).
31. W. B. Davenport, Jr., and W. L. Root, *An Introduction to the Theory of Random Signals and Noise*, pp. 96, 373. McGraw-Hill, New York, 1958.
32. R. B. Frieden, "Band-Unlimited Reconstruction of Optical Objects and Spectra," *J. Opt. Soc. Am.* **57,** No. 8, 1013-1019 (August 1967).
33. C. W. Barnes, "Object Restoration in a Diffraction Limited Imaging System," *J. Opt. Soc. Am.* **56,** 575 (1966).
34. H. A. Brown, "Effect of Truncation on Image Enhancement by Prolate Spheroidal Functions," *J. Opt. Soc. Am.* **59,** No. 2, 228-229 (February 1969).
35. N. J. Bershad, "Resolution, Optical-Channel Capacity and Information Theory," *J. Opt. Soc. Am.* **59,** No. 2, 157-163 (February 1969).
36. H. J. Landau and H. O. Pollak, "Prolate Spheroidal Wave Functions Fourier Analysis and Uncertainty—III," *BSTJ* **41,** No. 4 (July 1962).
37. D. A. Landgrebe and G. R. Cooper "Two-Dimensional Signal Representation Using Prolate Spheroidal Functions," *IEEE Trans. Commun. Theory* 30-40 (March 1963).
38. A. Papoulis, *Systems and Transforms with Applications in Optics*, p. 422. McGraw-Hill, New York, 1968.

Chapter 3 DIGITAL OPTICAL PROCESSING

3.0 IMPLEMENTATION

From the two-dimensional linear systems theory background of Chapter 2, it is evident that there are two mathematical techniques avilable for implementation of spatial filtering principles—convolution and Fourier transformations. In a computer environment, convolution is equivalent to vector-circulant matrix multiplication, a technique which for one-dimensional signals requires $N(N - 1)$ operations where N is the dimension of the data vector. The Fourier technique, using a fast transform algorithm, results in $2N \log N + N$ operations for the equivalent Fourier transform, scalar multiplication, and inverse transform operation. However, when not all degrees of freedom in the definition of the filter function are necessary, convolution techniques can be faster [1]. Yet to fully utilize the power of the computer approach, all degrees of freedom will become necessary, especially in matched filtering, and therefore the Fourier approach will be pursued.

Since the operation performed by a lens in the coherent optical system can be described by a Fourier transform equation, it is possible to simulate optical processing on a digital computer by evaluating the two-dimensional Fourier transform of a function mathematically. If $f(x, y)$ is a two-dimensional arrary of points of dimension N by N, the two-dimensional Fourier transform, $F(u, v)$, in discrete form, is defined as

$$F(u, v) = \frac{1}{N} \sum_{x=0}^{N-1} \sum_{y=0}^{N-1} f(x, y) \exp\left\{\frac{-2\pi i}{N}(xu + yv)\right\} \quad (1)$$

The evaluation of Eq. (1) can be broken into two steps. First, the one-dimensional Fourier transform $F(u, y)$, is taken along the x-coordinate of every horizontal line of $f(x, y)$. Then a second one-dimensional Fourier transform is taken in the y-direction of every vertical line of $F(u, y)$ to yield the composite transform, $F(u, v)$.

A filter function, $H(u, v)$, represented over a square plane of N^2 points is multiplied point-by-point with $F(u, v)$ to perform a filtering operation. Then the discrete two-dimensional Fourier transform of the product of $F(u, v)$ and $H(u, v)$ yields the filtered version of $f(x, y)$. Thus,

$$g(x, y) = \frac{1}{N} \sum_{u=0}^{N-1} \sum_{v=0}^{N-1} F(u, v) H(u, v) \exp\left\{\frac{-2\pi i}{N}(ux + vy)\right\} \quad (2)$$

is a simulated representation of the output of a coherent optical processing system with filter function $h(x, y)$.

Until very recently, only brute force methods were available for producing Fourier transforms digitally. By conventional techniques, the evaluation of Eq. (1) would require $2N^3$ complex multiplications and additions. Fortunately, algorithms have been developed which

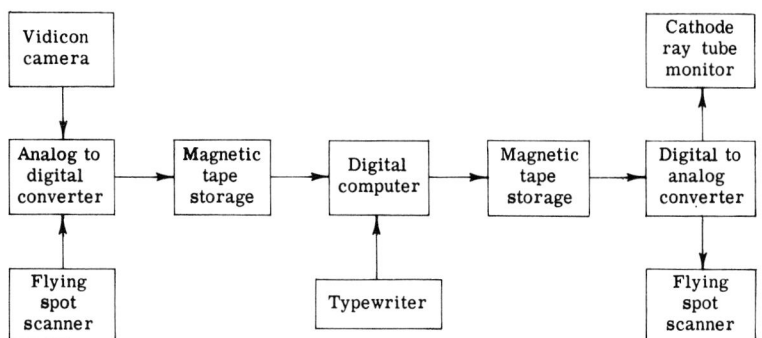

Fig. 1. Computer image processing equipment.

require only $2N^2 \log N$ complex additions and complex multiplications using the above two-step technique. Such algorithmic simplifications will be described later. With these algorithms, two-dimensional Fourier transforms of images with 256 by 256 points can be performed by a digital computer in from 5 to 20 minutes depending upon the speed and memory capacity of the computer.

Figure 1 is a block diagram of a digital optical processing system. Photographs or transparencies of images are scanned by either a vidicon camera or flying spot scanner and converted to digital form for storage on magnetic tape. The computer takes the two-dimensional Fourier transform of the digitized image, multiplies the transformed image by a digitally stored transfer function, and then takes a second Fourier transform of the product. This digitally processed image is then converted to analog form for display and photographing on a cathode-ray tube monitor. Figure 2 displays a test scene for verification of the computer technique. The original image, the block letters "USC," appears in Fig. 2a. The magnitude of the Fourier transform appears in Fig. 2b. The magnitude, or some property of the magnitude, must be displayed due to the inability of optical devices to sense other than intensity information. It is evident from Figs. 2(b) and 2(c) that the majority of the spectral energy is concentrated along the horizontal and vertical axes, which is indeed the case due to the edge orientations of the block letters. A second Fourier transform results in the image of Fig. 2(d) which is the rotated view of the original. The rotation is the result of Eq. (10i) in Chapter 2. In future examples, the double Fourier transforms will be reoriented for better viewing.

Figure 3 displays some interesting "zero-one" filter results for two block letter test scenes [2]. Figure 3(a) is the result of two-dimensional low pass spatial filtering, whereas Fig. 3(b) is the result of two-dimensional high pass spatial filtering, and Fig. 3(d) is a combination of horizontal low pass and vertical high pass spatial filtering. It is immediately obvious that the digital computer is going to offer a spatial frequency domain where the traditional drawbacks of scaling, noise, resolution, and photographic nonlinearities of conventional coherent opitical data processing techniques have been eliminated.

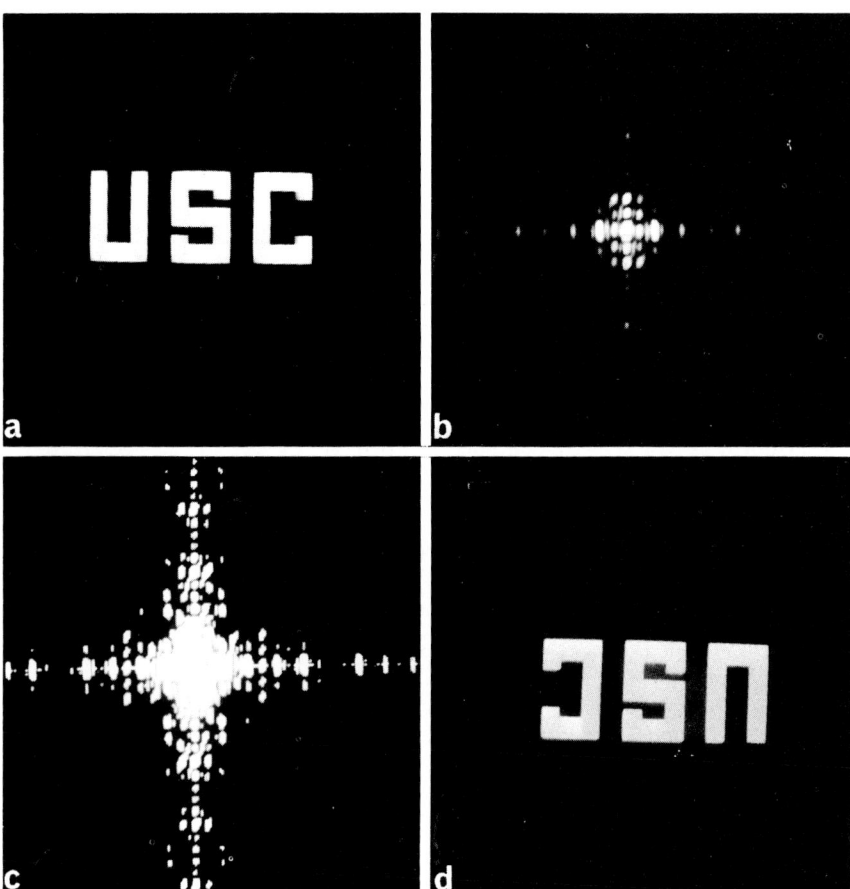

Fig. 2. Fourier transform of block letters. (a) Block letters. (b) Magnitude of Fourier transform. (c) Logarithm of magnitude of Fourier transform. (d) Double Fourier transform.

3.0 IMPLEMENTATION 35

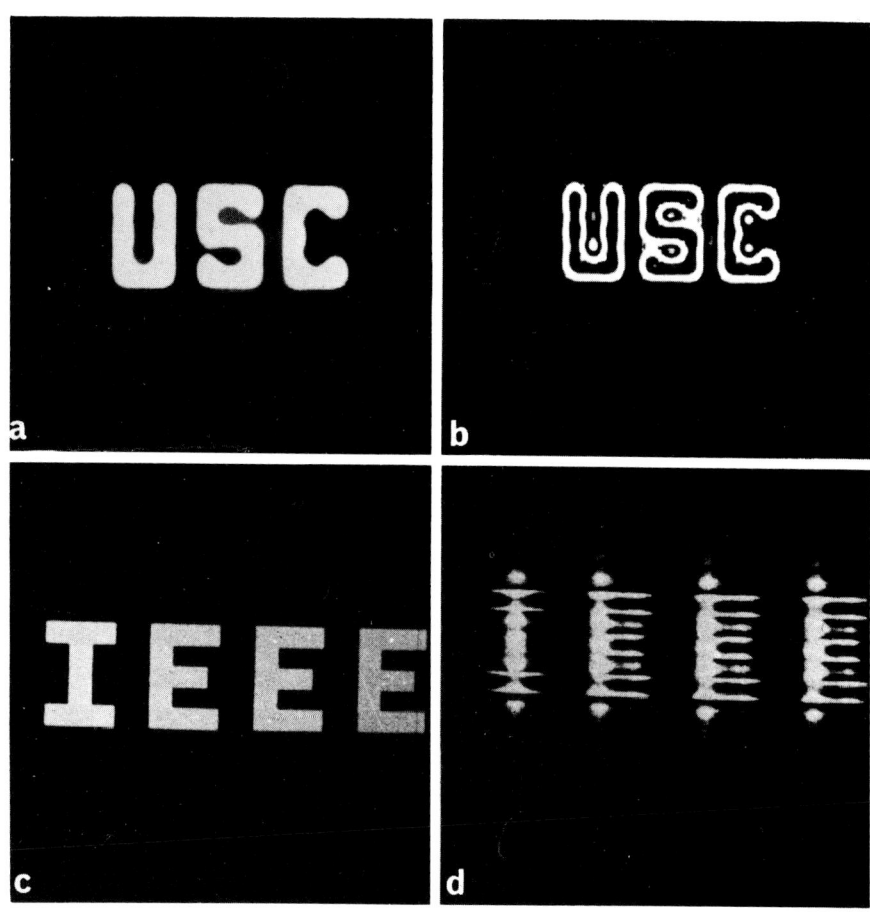

Fig. 3. "Zero-one" filtering. (a) Low pass "USC." (b) High pass "USC." (c) Original "IEEE." (d) Horizontal low pass, vertical high pass "IEEE."

3.1 IMAGE ENHANCEMENT

It was indicated in the previous chapter that there exist two techniques to improve images. The first and most logical is to approximate the OTF by its inverse and thereby perform inverse filtering. Yet the best one can hope to do using inverse filtering is to achieve the ideal low pass system offered by purely diffraction limited optics. The next technique to improve images especially beyond the diffraction limit is to apply the mathematical principles of super resolution and analytic continuation. While research is still in the early stages for digital implementation of such techniques, initial results are quite encouraging. The physical limitations of large current and flux requirements for microwave and optical systems respectively can be easily handled with the simple expedient of floating point arithmetic to any precision necessary.

However, considerable image improvement can be obtained without resorting to super resolution techniques. Using inverse filtering principles in the Fourier transform domain of an image is equivalent to multiplication and in the digital computer all degrees of freedom are possible. However, it was mentioned in the last chapter that certain image degrading optical transfer functions can become negative, thereby possessing some zeros which cannot be inverted. The example of badly out-of-focus circularly symmetric imaging system equation (26c) of Chapter 2 has an optical transfer function given by that of Fig. 4(a) and a nonrealizable inverse filter of Fig. 4(b). When the OTF becomes negative, it causes a contrast reversal which is easily demonstrated by the crude test chart of Fig. 4(c) [3]. As the sectors come closer to the center, they define higher and higher spatial frequencies, thus resulting in the contrast reversal of Fig. 4(d). Upon close examination it will become evident that a second reversal occurs close to the center thus defining the second positive lobe in the transfer function of Fig. 4(a). It is, of course, a simple matter to digitally compensate for such an OTF by a realizable approximation to the nonrealizable inverse filter of Fig. 4(b).

The most obvious technique for inverse filtering when little is known about the OTF of the system producing the image is to assume a Gaussian transfer function which is quite realistic for many disturb-

3.1 IMAGE ENHANCEMENT 37

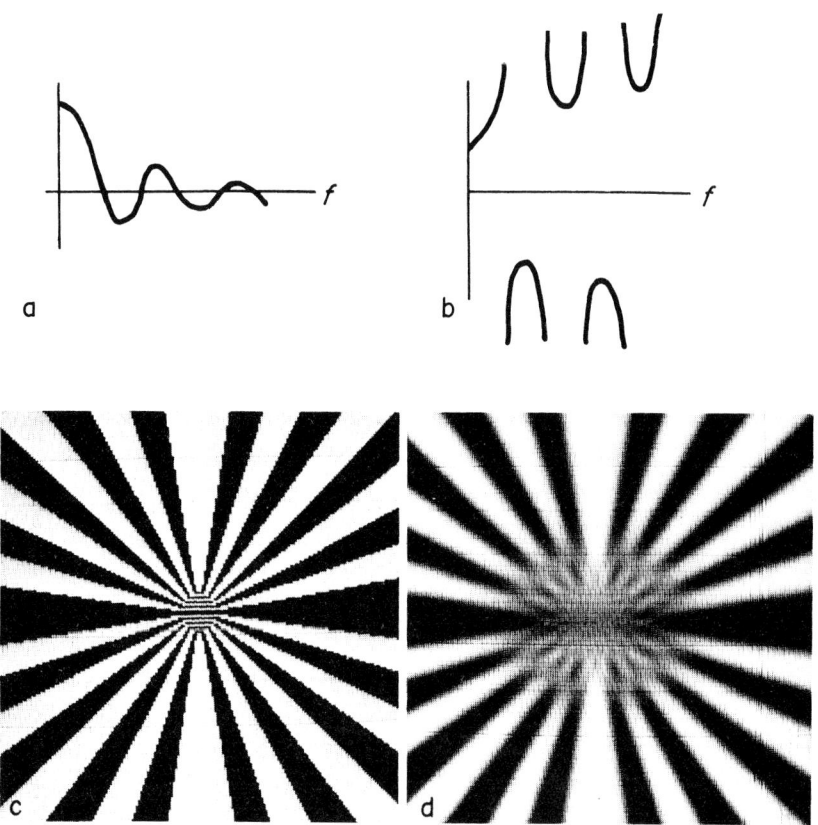

Fig. 4. Contrast reversal. (a) OFT given by $J_1(a\omega)/a\omega$. (b) Inverse filter. (c) Object before degradation. (d) Object after degradation.

38 DIGITAL OPTICAL PROCESSING

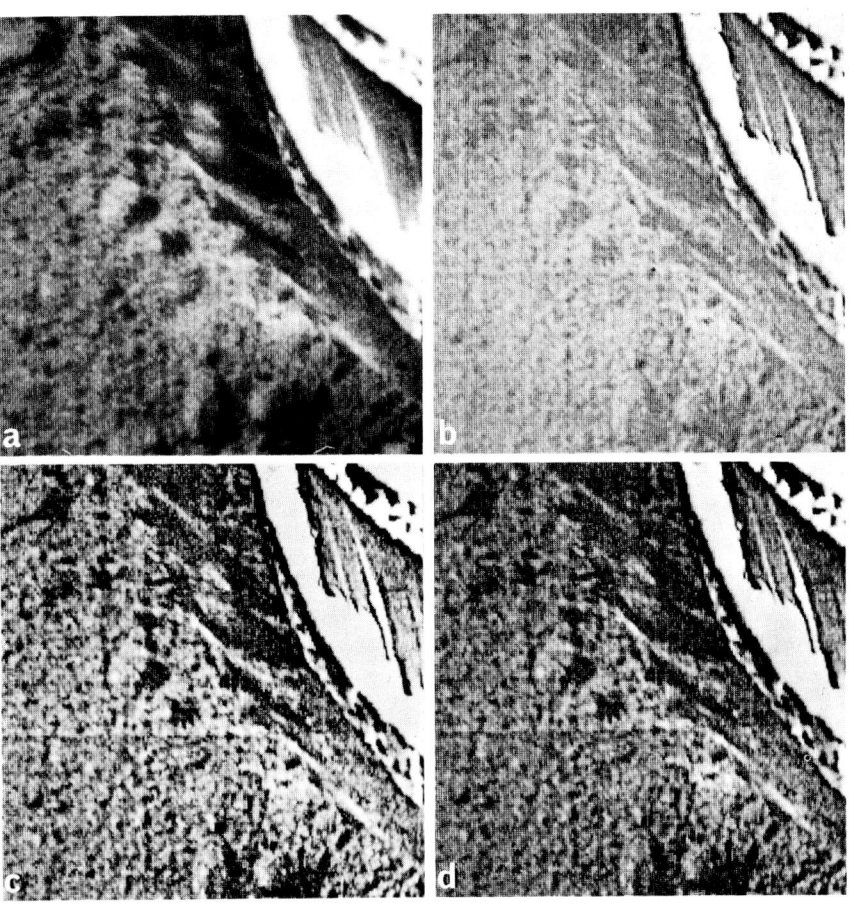

Fig. 5. Inverse Gaussian filter footpad. (a) Before filtering. (b) After filtering, no clipping. (c) After filtering, 25% clipping. (d) After filtering, 50% clipping.

3.1 IMAGE ENHANCEMENT 39

Fig. 6. Inverse Gaussian filter. (a) Test chart before filtering. (b) Test chart after filtering. (c) Moon surface before filtering. (d) Moon surface after filtering.

ing phenomena [4]. The compensating filter then becomes an inverse Gaussian curve which is easily implemented on a digital computer. Assuming no *a priori* knowledge about one axis being more sensitive than another, the inverse filter will then be circularly symmetric and defined by one parameter, the variance of the curve. The parameter can be adjusted as suggested in Chapter 2 possibly in a man-machine interactive mode until a maximum or pleasing contrast in filtered image is obtained. This is the technique which has been utilized in obtaining the results of Figs. 5 and 6. Figure 5 presents before and after scenes obtained from the moon's surface via the Surveyor spacecraft. An increased contrast is particularly evident in the footpad scene of Figs. 5(c) and 5(d). Also the high frequency information in the footpad has been improved. This is quite evident in the fine structure within the footpad itself. Notice the enhanced but non-clipped image of Fig. 5(b). It must be remembered that in the computer environment one will obtain "filtered" images that have an increased range of definition. Thus the filtered footpad of Fig. 5(b) has both negative and positive numbers so that for display purposes it is necessary to add a constant value to guarantee a positive picture which then must be renormalized. This results in an enhanced image with poor contrast because we have normalized beeween the peaks of "ringing" edges instead of clipping away such peaks. The clipping procedure results in the much more appealing images of Figs. 5(c) and 5(d). Figure 6 presents two different scenes which have been inverse Gaussian filtered and clipped for pleasing contrast. Notice the greater detail in the area to the left of the test chart in Fig. 6(b). Also notice the very high frequency noise which has become viewable in the test chart itself. In the filtered moon surface of Fig. 6d, there is a general overall improvement in focus or definition with particularly pleasing results in the rock in the lower left corner of the enhanced image.

3.2 COMPUTER CALCULATED DIFFRACTION PATTERNS

Knowledge of aperture diffraction patterns is useful in the design of optical systems and the prediction of antenna preformance. For complex apertures, hand computation often is not feasible, and optical models are difficult to construct [5]. The diffraction pattern can be quickly calculated to a high degree of accuracy with a digital computer [6].

With the computer method the electric field amplitude and phase in the aperture plane is specified over an array of N by N elements. A two-dimensional Fourier transform is taken of the aperture array yielding the spatial diffraction pattern. The magnitude squared of each Fourier sample is formed to give the conventional intensity or power pattern, which may then be displayed on a CRT for photographic recording. Since the computer can calculate with arbitrarily high precision, low intensity data, normally lost in an optical system display, can be maintained with the computer technique. One means of doing this to display the logarithm of the diffraction pattern intensity to enhance the low intensity data. Other linear and nonlinear weightings may be applied to investigate particular areas of interest in the diffraction pattern. In addition to displaying the diffraction pattern by intensity variations on a CRT, a three-dimensional perspective view can be displayed by a graphical display program. In this manner the relative amplitudes of points in the diffraction pattern are readily compared.

The photographs of Fig. 7 illustrate the aperture intensity, magnitude of the diffraction pattern intensity, and the logarithm of the diffraction pattern intensity for a uniformly illuminated square aperture. Both the aperture and diffraction pattern display contain 256 by 256 points. The analytic solution to the diffraction pattern of a square is $[(\text{sinc } au)(\text{sinc } av)]^2$. The logarithmic display of the diffraction pattern of the square shown in Fig. 7(c) vividly brings out the data lost in the normal display. Figure 8 contains a uniformly illuminated circular aperture, its diffraction pattern, and a perspective view of the diffraction pattern. The analytic solution to this pattern is a first-order Bessel function divided by its argument, $[J_1(a\omega)/a\omega]^2$, and in classical optics is known as an Airy disk. The fact that the perspective display is not circularly symmetric is because the original aperture was a circle approximated by a rectangular grid which defined some nonsymmetric high frequency data. Figure 9 is the diffraction pattern of an unsymmetric cross. The logarithmic display vividly brings out the information missed in the magnitude display. This particular

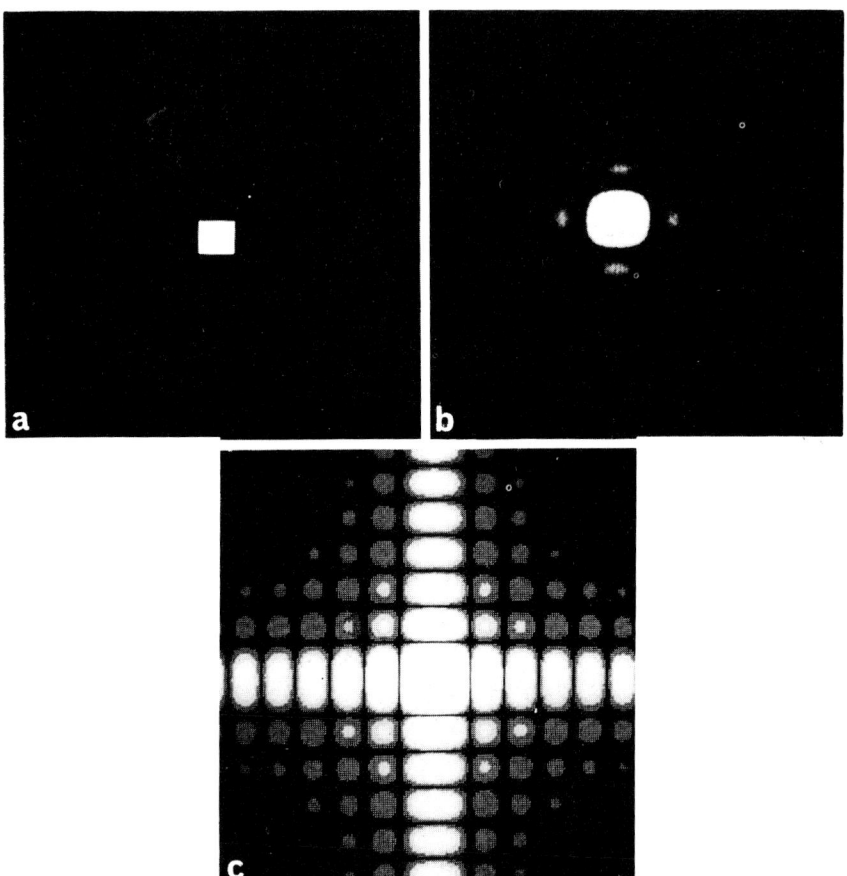

Fig. 7. Diffraction pattern of a square. (a) Square aperture. (b) Diffraction pattern. (c) Logarithm of diffraction pattern.

example exemplifies the power of the computer technique since, even though the analytic solution can be obtained from a superposition of the sinc functions resulting from the rectangles defining the cross, the shape and location of the zeros and the main beam will still be difficult if not impossible to visualize. In addition the computer technique provides a plane of data in which the main beam to side lobe energy ratio can be easily calculated. The computer technique can also be utilized as an analysis tool for measuring the side lobe energy losses due to various geometrical interference object in the aperture plane. Figure 10 illustrates such a simulation in which a uniformly illuminated circular aperture is compared with the same aperture with feed-horn struts simulated by the cross hairs. The energy loss is evident in the comparison of Figs. 10(b) and 10(d). Various aperture shading techniques can also be simulated on the computer. Figure 11 is one such example. In this case a uniformly illuminated square aperture is compared with a one-dimensional inverse parabolic illuminated aperture. It is evident from Fig. 11(d) that for the nonuniform illumination the energy in one-dimension is compacted and spread out in the other dimension. Arrays of apertures and interference patterns can also be investigated using the computer. Figure 12 is one such example in which four deterministically placed arrays of uniformly illuminated square antennas are investigated for their far field Fraunhofer diffraction patterns. This example vividly demonstrates the linearity and superposition of the Fourier operation. The envelop of the diffraction pattern is determined by the Fourier transform of one of the squares in the aperture plane. The resulting constructive and destructive interference modulation under the envelop is due to the phases of the individual diffraction patterns of each of the four squares in the aperture plane. The physical relationship of the four apertures to one another defines the deterministic crosshatching modulation in the diffraction plane. This experiment was carried out in an optical environment by Michaelson in the early part of the century [7]. The last example of computer calculated antenna diraction patterns is that given by Fig. 13 in which a uniformly illuminated isosceles triangular aperture is Fourier transformed to obtain its diffraction pattern. Note in this case as well as in Fig. 8, the space quantization of the definition of the triangle introduces high frequency

44 DIGITAL OPTICAL PROCESSING

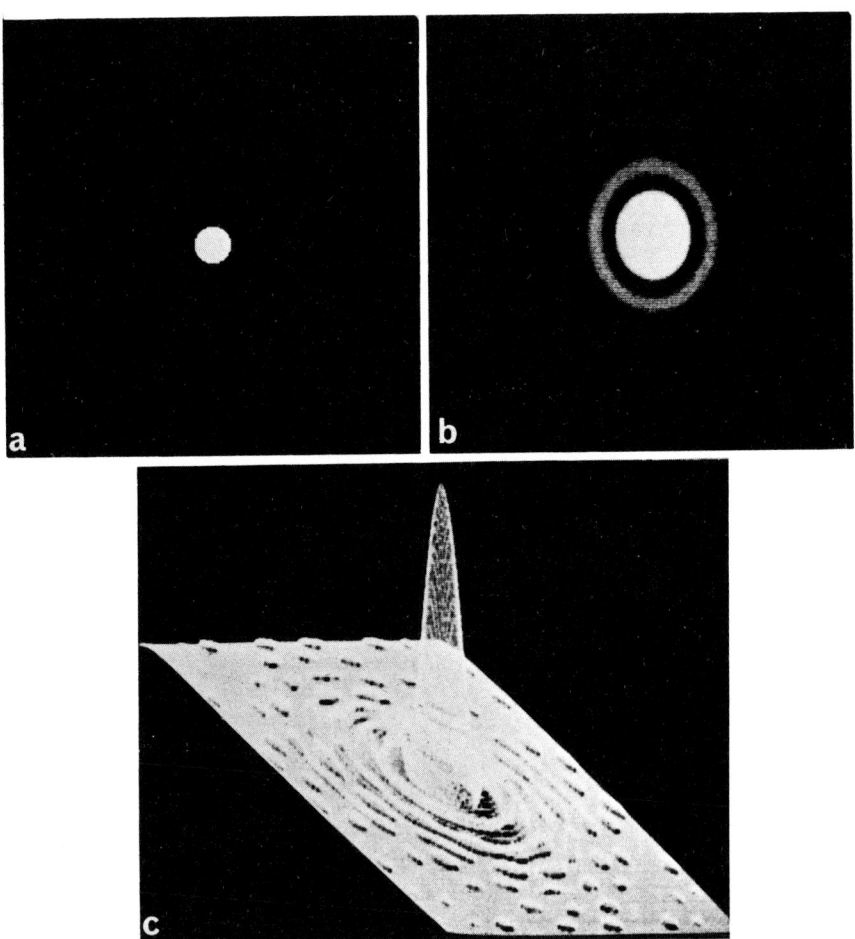

Fig. 8. Diffraction pattern of a circle. (a) Circular aperture. (b) Diffraction pattern. (c) Perspective of diffraction pattern.

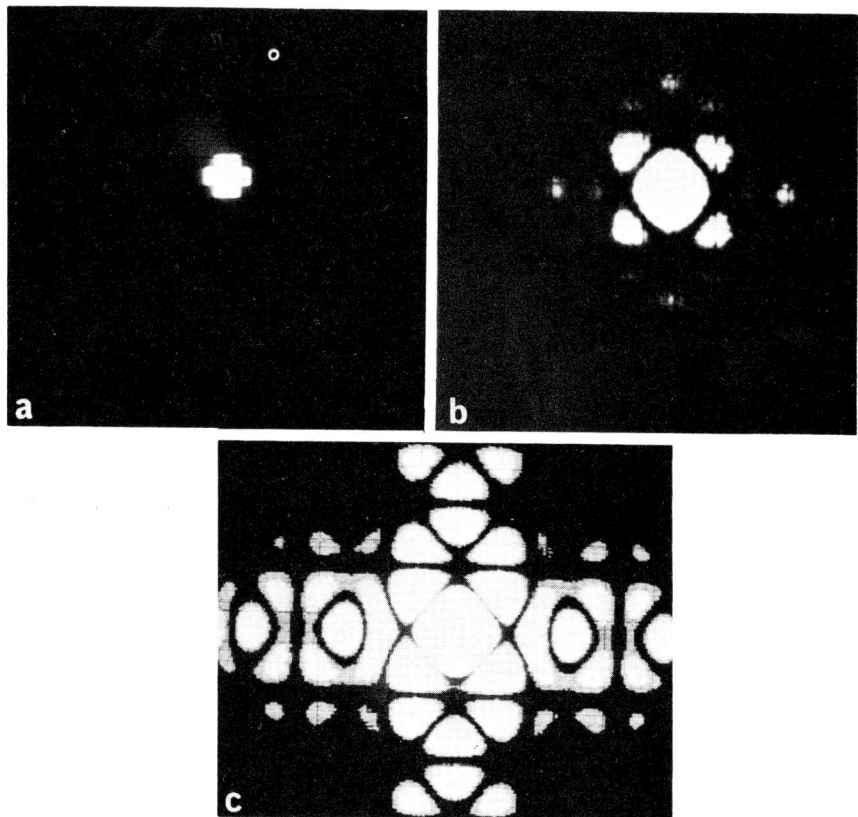

Fig. 9. Diffraction pattern of an unsymmetrical cross. (a) Aperture. (b) Diffraction pattern. (c) Logarithm of diffraction pattern.

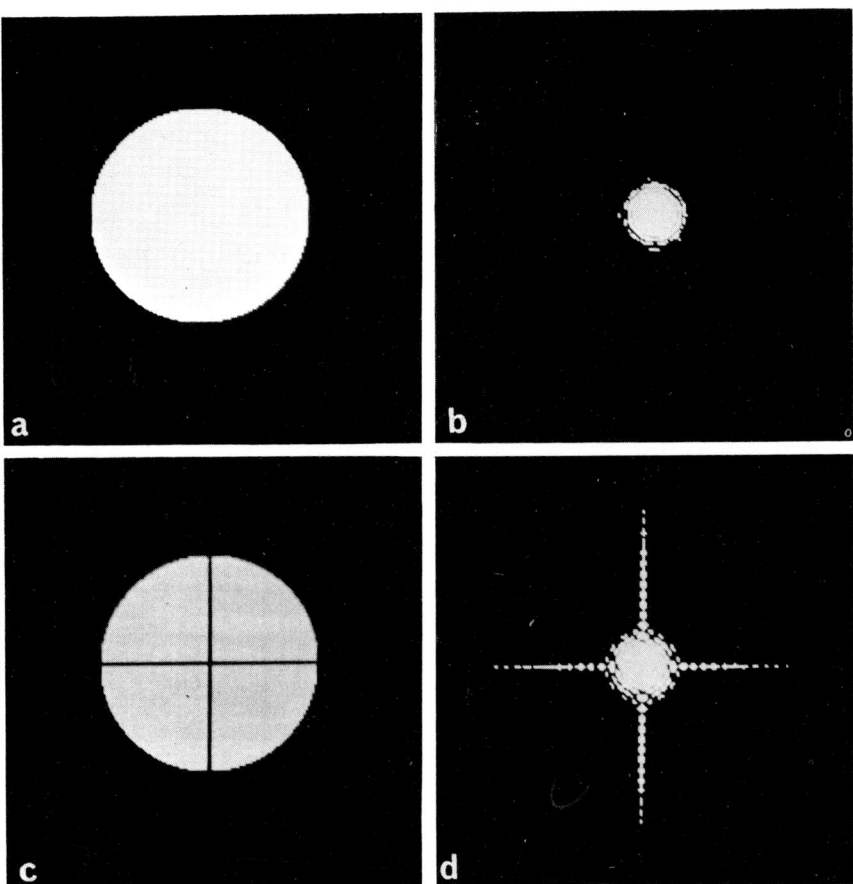

Fig. 10. Diffraction patterns of geometrical obstructions. (a) Circular aperture. (b) Diffraction pattern. (c) Blocked circular aperture. (d) Diffraction pattern.

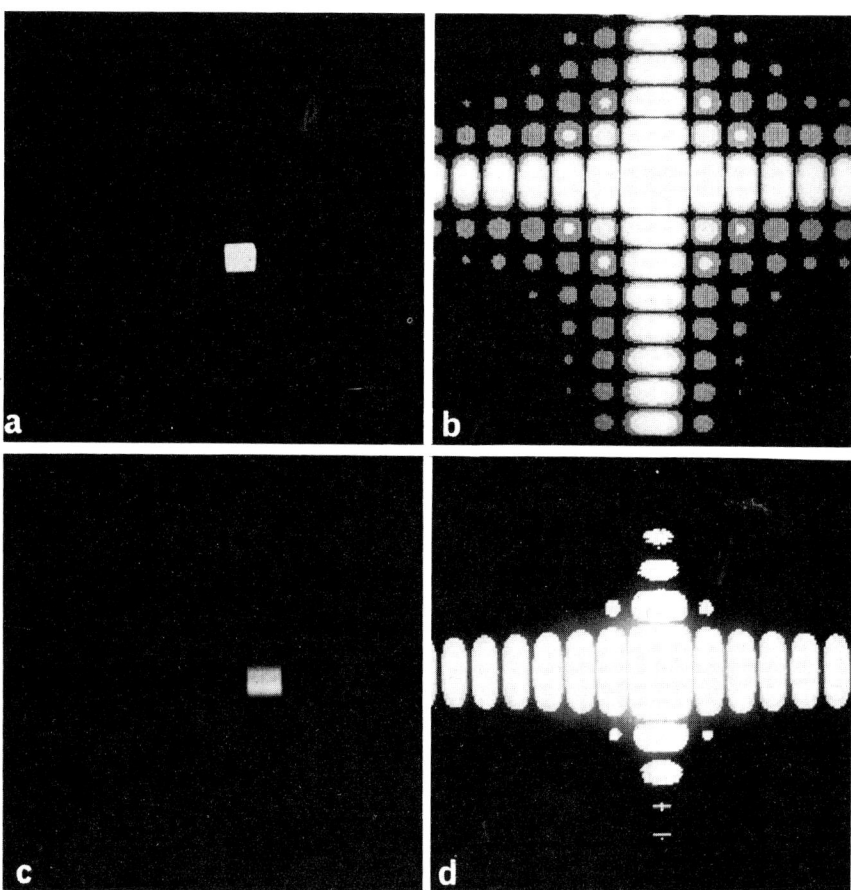

Fig. 11. Aperture illuminations. (a) Uniform illumination. (b) Logarithm of diffraction pattern. (c) One-dimensional inverse parabolic illumination. (d) Logarithm of diffraction pattern.

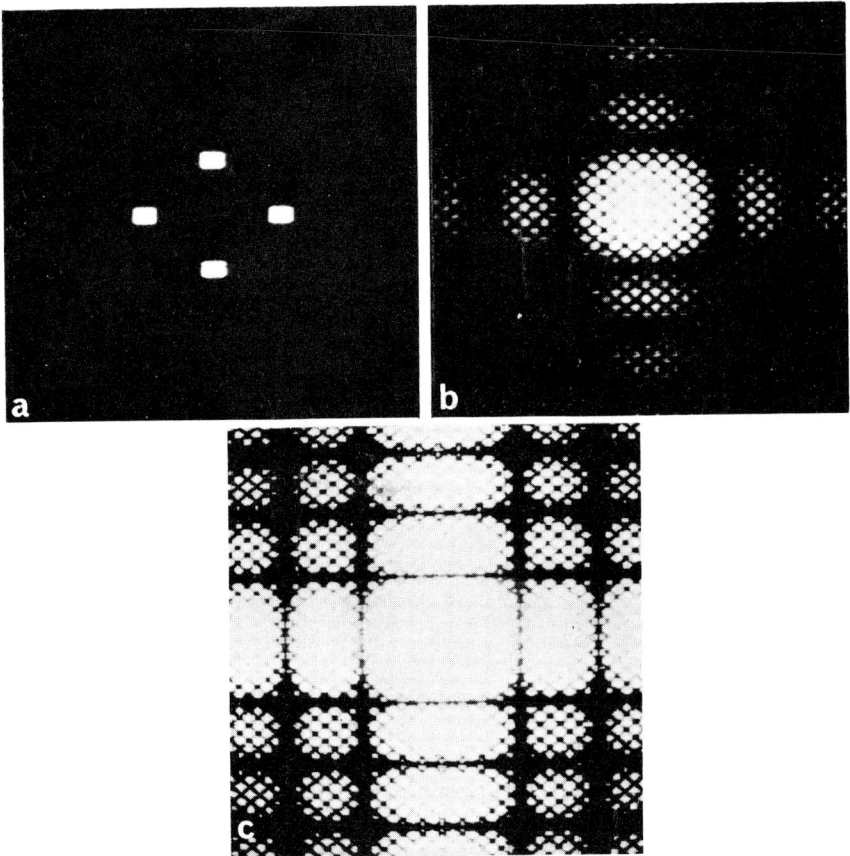

Fig. 12. Array patterns. (a) Four square array. (b) Diffraction pattern. (c) Logarithm of diffraction pattern.

3.2 COMPUTER CALCULATED DIFFRACTION PATTERNS

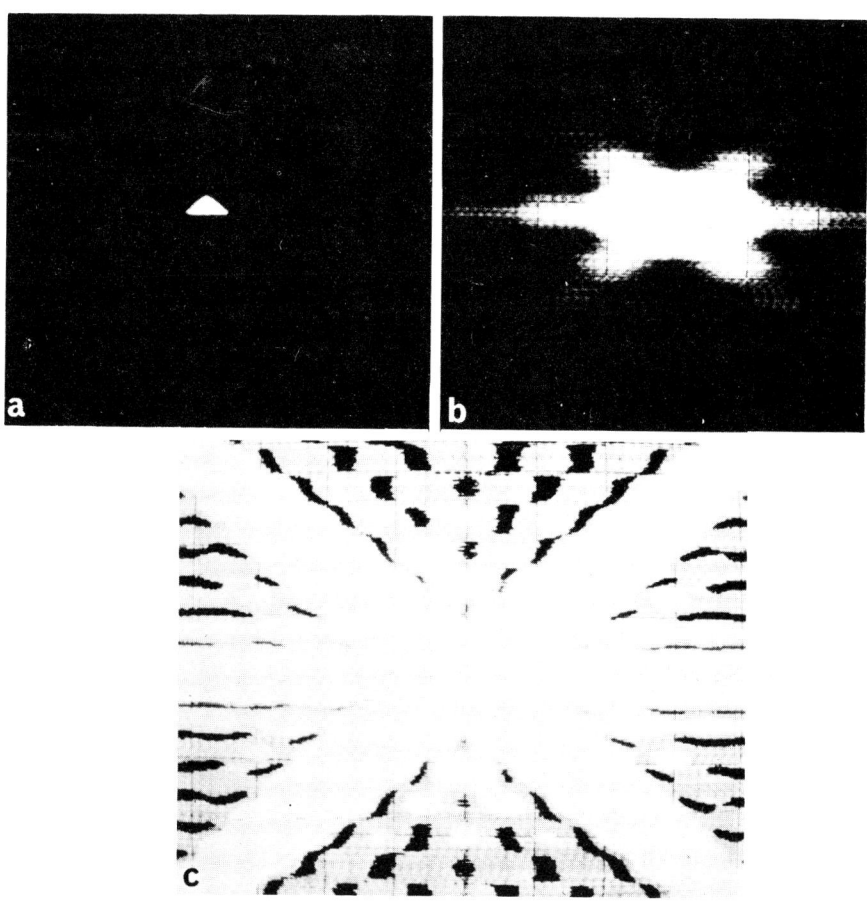

Fig. 13. Triangular aperture. (a) Aperture. (b) Diffraction pattern. (c) Logarithm of diffraction pattern.

energy which normally would not be present. This high frequency effect can be reduced by taking higher dimensional transformations with larger triangular and therefore finer spatial quantization.

3.3 DIGITAL FOURIER HOLOGRAPHY

An optical hologram records the spatial amplitude and phase variations over the surface of a photographic film or other suitable recording device. A special class of holograms, known as Fourier holograms, essentially stores the two-dimensional Fourier transform of an object shifted by a spatial carrier frequency [8]. Let $f(x, y)$ be the two-dimensional electric field representation of an object and let the Fourier transform of $f(x, y)$ be written in magnitude-phase form as

$$F(u, v) = |F(u, v)| \exp\{j\phi(u, v)\} \quad (3)$$

at a holographic plate and let $\exp\{j(\theta u + \Phi v)\}$ represent a uniform coherent light beam striking the plate at angles θ and Φ with respect to the normal. The hologram $I_n(u, v)$ then becomes

$$I_n(u, v) = |\exp\{j(\theta u + \Phi v)\} + |F(u, v)| \exp\{j\phi(u, v)\}|^2 \quad (4a)$$
$$I_n(u, v) = 1 + |F(u, v)|^2 + |F(u, v)| \cos(\phi(u, v) + \theta u + \Phi v) \quad (4b)$$

An improvement on this equation can be realized by computer implementation. The second term, $|F(u, v)|^2$, of Eq. (4) can be eliminated thereby allowing lower spatial carrier frequencies (θ, Φ) for reconstruction purposes. Thus for computer implementation the Fourier transform of $f(x, y)$ is taken by the computer and an intensity function, $I(u, v)$, is formed mathematically.

$$\begin{aligned} I(u, v) = &\tfrac{1}{2} F(u, v) \exp\{i(\theta u + \Phi v)\} \\ &+ \tfrac{1}{2} F^*(u, v) \exp\{-i(\theta u + \Phi v)\} + A \end{aligned} \quad (5)$$

where A is the maximum value of $|F(u, v)|$. Equation (5) can also be written as

$$I(u, v) = |F(u, v)| \cos\{\theta u + \Phi v + \phi(u, v)\} + A \quad (6)$$

indicating that $I(u, v)$ is a positive, real function containing only the subject spectrum of interest. The digital reconstruction process consists simply of taking the Fourier transform of $I(u, v)$. Then, by the Fourier transform shifting theorem

$$\mathscr{F}\{I(u, v)\} = f(x - \theta, y - \Phi) \\ + f(-x - \theta, -y - \Phi) + A\delta(x, y) \quad (7)$$

where $A\delta(x, y)$ represents a bright spot in the center of the reconstruction plane which can be eliminated if desired. The resulting output consists of the original $f(x, y)$ spatially shifted by θ and Φ, and a 180° rotated version of the original $f(-x, -y)$ shifted in the oppo-

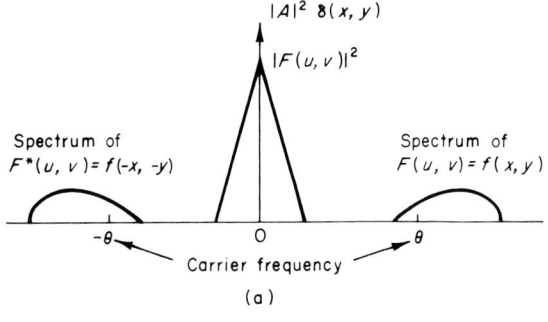

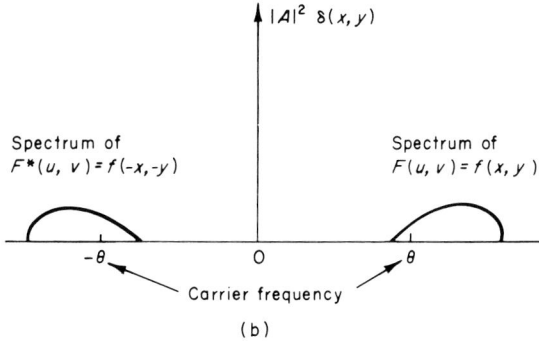

Fig. 14. Hologram spectra. (a) One-dimensional interpretation of spectrum of optically constructed holograms. (b) One-dimensional interpretation of spectrum of computer holograms.

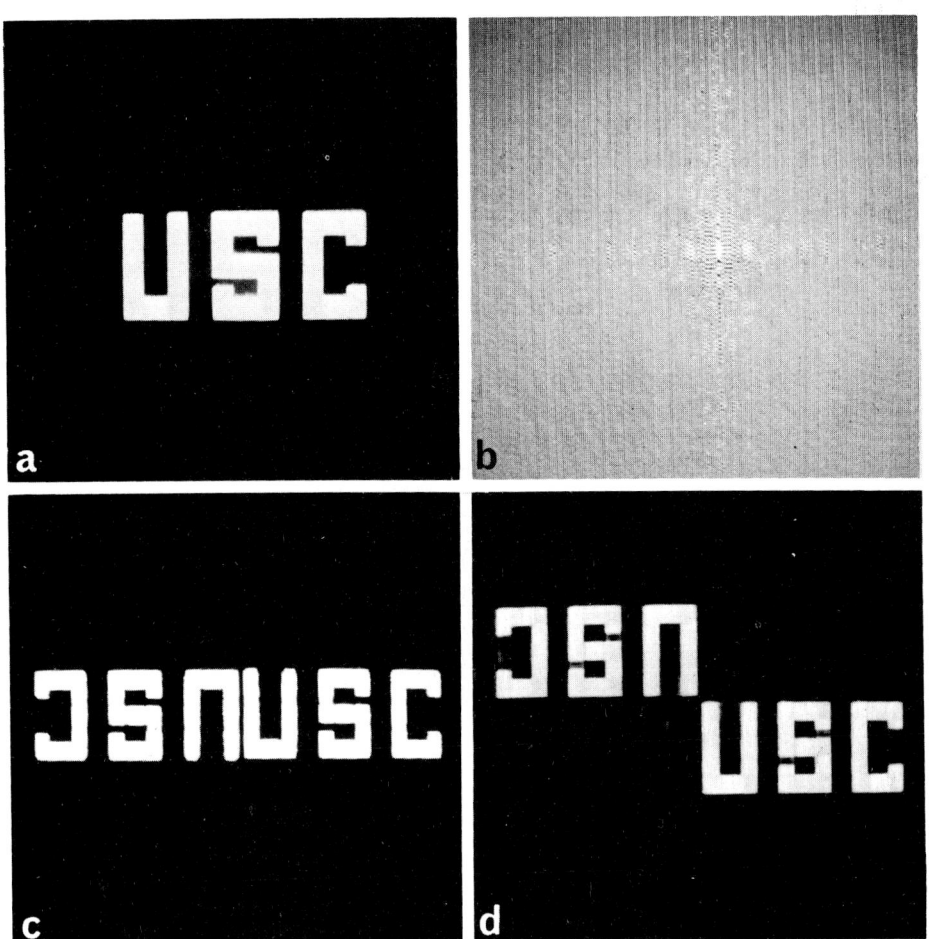

Fig. 15. Digitally generated and reconstructed Fourier holograms. (a) Original scene. (b) Digital hologram of original. (c) Reconstruction: one-dimensional carrier. (d) Reconstruction: two-dimensional carrier.

site direction by θ and Φ. Because the $|F(u, v)|^2$ term of Eq. (4b) has been eliminated, the spatial carrier frequencies (θ, Φ) need only be chosen to insure no overlap of the $f(x, y)$ function with its rotated version, Eq. (7); see Fig. 14.

Figure 15 illustrates the construction and reconstruction of a digital Fourier hologram of the block letters "USC." The Fourier transform of the letters in Fig. 15(a), multiplied by spatial carrier, is shown in Fig. 15(b). This Fourier hologram contains both light and dark amplitude variations. Phase variations are represented as shifts in the underlying spatial carrier. Figure 15(c) is the digital reconstruction of the Fourier hologram of Fig. 15(b) showing the twin images for a one-dimensional spatial carrier. Figure 15(d) is the reconstruction for a two-dimensional carrier. Theoretically the same images Figs. 15(c) and 15(d), would result if the hologram of Fig. 15(b) were scaled and placed in an optical bench.

3.4 CONCLUSIONS

Chapter 3 has presented some results of computer implementation of the techniques and ideas suggested in Chapter 2. After a brief description of implementation procedures for a digital environment, test scenes were then presented as verification of the tools developed. As an example some unusual "zero-one" filters were implemented to give the reader a feeling for the concepts of spatial frequencies along given axes. The principle of inverse filtering for image enhancement purposes were then demonstrated initially by the out-of-focus optical transfer function which produces contrast reversals as demonstrated in Fig. 4. The computer was used to implement inverse Gaussian filter techniques and the resulting image enhanced figures were then presented. It is important to realize that the computer has made use of active filters, filter values greater than unity, whereas in most optical systems only passive filtering is possible. An immediate application of a two-dimensional linear systems implementation in a digital computer is the calculation of diffraction patterns. Examples were presented for traditional aperture illumination and shape as well as some nontraditional apertures. Finally the concepts of digital Fourier holography were developed, and it was shown for computer generated holograms more efficient utilization of the spatial frequency carrier modulation concepts could be developed. This chapter was presented in order to demonstrate some of power and capability of digital implementation of optical processes. The results presented here are not to be implied as constructing the entire spectrum of optical processing and as we shall see in further chapters, the computer will be quite useful in other optical as well as nonoptical image processing situations.

REFERENCES

1. R. H. Selzer, "The Use of Computers to Improve Biomedical Image Quality," *AFIPS 1968 Fall Joint Computer Conf.* **33**, Pt. 1, pp. 817-834.
2. H. C. Andrews and W. K. Pratt, "Digital Computer Simulation of Coherent Optical Processing Operations," *IEEE Computer Group News* **2**, No. 6, 14-19 (1968).
3. J. W. Goodman, *Introduction to Fourier Optics* p. 124. McGraw-Hill, New York, New York (1968).
4. P. F. Mueller and G. O. Reynolds, "Image Restoration by Removal of Random-Media Degradation," *J. Opt. Soc. Am.* **57**, No. 11, 1338-1344 (November 1967).
5. A. L. Ingalls, "Optical Simulation of Microwave Antennas," *IEEE Trans. Antennas Propagation* **AP-14**, No. 1, 2-6 (January 1966).
6. W. K. Pratt and H. C. Andrews, "Computer Calculated Diffraction Patterns," *Appl. Opt. Letters* **7**, No. 2, 378-379 (February 1968).
7. A. A. Michelson, *Studies in Optics*, p. 70. Chicago Univ. Press, Chicago 1927.
8. J. B. DeVelis and G. O. Reynolds, *Theory and Applications of Holography*. Addison-Wesley, Reading, Massachusetts, 1967.

Chapter 4 TWO-DIMENSIONAL MATCHED FILTERING

4.0 INTRODUCTION

The concept of filtering in two dimensions has been introduced in the study of image enhancement in which a filter was selected to provide some image improvement function. The filtering concept fell nicely into the context of two-dimensional linear systems and was a simple extension of one-dimensional linear filter theory. A different type of filter, known as the two-dimensional matched filter, can also be described as an extension of one-dimentional filtering theory. However, the most unique characteristic of the two-dimensional matched filter is that it is not used to improve images but is used to detect two-dimensional functions. In the process of image detection, the image information is destroyed in the sense of visually pleasing results and the output of the filter is no longer an image but a correlation plane whose amplitude at a given coordinate corresponds to the degree of correlation of the input image with the desired image, known as the signal. The matched filter results in the optimum filter for detection purposes in the sense of signal-to-noise ratio, likelihood ratio, and inverse probability criteria. The filter is often implemented as a correlation receiver in radar and communication systems and has applications in the detection of images, gradient images, and in the use of image evaluation.

4.1 THEORY

The matched filter has been developed historically as a one-dimensional detection filter for various communication systems [1-4]. The filter is optimum in maximizing the signal-to-noise ratio and can be derived on the basis of three different criteria. The three criteria are:

(1) signal-to-noise;
(2) likelihood ratio;
(3) inverse probability.

An excellent treatment of all three is presented in Cook and Bernfeld [5]. In the one-dimensional case the matched filter impulse response for a signal, $f(x)$, in the presence of white noise is given by

$$h(x) = f^*(\xi - x) \tag{1a}$$

with system response given by

$$G(u) = F^*(u)\exp\{-j\xi u\} \tag{1b}$$

The two-dimensional matched filter can be easily derived using the signal-to-noise ratio criteria, S/N, in the following way. Let $f_I(x,y)$ be the input signal to be detected in the presence of noise $n(x,y)$, $g(x,y) = f_I(x,y) + n(x,y)$. Assume the noise process is wide sense stationary, ergodic in autocorrelation, $R_N(\tau, T)$, with an input power spectrum $S_{N_I}(u,v) = \mathscr{F}\{R_N(\tau, T)\}$. Then the filter output noise power spectrum is given by

$$S_{N_O}(u,v) = S_{N_I}(u,v)|H(u,v)|^2 \tag{2a}$$

with output noise power

$$N = \iint_{-\infty}^{\infty} S_{N_O}(u,v)\, du\, dv \tag{2b}$$

The filter output, $f_O(x,y)$, of the signal is given by the convolution, $f_I(x,y) \circledast h(x,y)$. The instantaneous signal energy, S, is given by

$$S = |f_O(\xi, \eta)|^2 \tag{3a}$$

$$= \left|\iint_{-\infty}^{\infty} F_O(u,v)\exp\{j(u\xi + v\eta)\}\, du\, dv\right|^2 \tag{3b}$$

$$= \left| \iint_{-\infty}^{\infty} F_I(u,v) H(u,v) \exp\{j(u\xi + v\eta)\} \, du \, dv \right|^2 \quad (3c)$$

The signal-to-noise ratio assuming white noise,

$$S_{N_I}(u,v) = \frac{N_O}{2},$$

is given by

$$\frac{S}{N} = \frac{2 \left| \iint_{-\infty}^{\infty} F_I(u,v) H(u,v) \exp\{j(u\xi + v\eta)\} \, du \, dv \right|^2}{N_O \iint_{-\infty}^{\infty} |H(u,v)|^2 \, du \, dv} \quad (4a)$$

and by the Schwartz inequality

$$\frac{S}{N} \leq \frac{2 \iint_{-\infty}^{\infty} |F_I(u,v)|^2 \, du \, dv \iint_{-\infty}^{\infty} |H(u,v)|^2 \, du \, dv}{N_O \iint_{-\infty}^{\infty} |H(u,v)|^2 \, du \, dv} \quad (4b)$$

$$= \frac{2}{N_O} \iint_{-\infty}^{\infty} |F_I(u,v)|^2 \, du \, dv \quad (4c)$$

Equality between Eqs. (4a) and (4b) or (4c) can be seen to hold for the system response

$$H(u,v) = F_I^*(u,v) \exp\{-j(u\xi + v\eta)\} \quad (5a)$$

resulting in an impulse response of

$$h(x,y) = f_I^*(\xi - x, \eta - y) \quad (5b)$$

For the case of nonwhite noise the matched filter becomes [6–8]

$$H(u,v) = \frac{F_I^*(u,v) \exp\{-j(u\xi + v\eta)\}}{S_{N_I}(u,v)} \quad (5c)$$

which is equivalent to the white noise matched filter preceded by a whitening filter of system response proportional to the inverse of the noise power spectrum. The impulse response of this filter can be viewed as

$$h(x, y) = f_I^*(\xi - x, \eta - y) \circledast Q_N(x, y) \qquad (5d)$$

where

$$Q_N(x, y) = \mathscr{F}^{-1}\{[S_{N_I}(u, v)]^{-1}\}$$

One interpretation of the operation of the matched filter is to convolve the impulse response of the filter with the input signal to the system. However, if the input signal is the signal to be detected, a perfect match will occur when the signal and filter (which is the conjugate of the signal now rotated a total of 360°) identically overlap each other. This will occur at the coordinates (ξ, η) which then describe the position of the detected two-dimensional signal. Thus, due to the convolution operation of linear systems, the matched filter becomes translation invariant because convolving the input with the filter response searches the entire plane for a possible match. Consequently, the filter can be used to detect multiple targets spaced at various translations in the input plane,

$$\sum_i f_I(x - \xi_i, y - \eta_i).$$

It should be mentioned that since the impulse response of the filter is a rotated and conjugated version of the signal to be detected, the operation of convolution actually results in a correlation of the input plane with the conjugate of the signal of interest. Thus the matched filter results in the correlation receiver implementation so common in one-dimensional detection systems. It is interesting to note that the two-dimensional matched filter has an added degree of detectability in the sense that it is invariant to translations in either direction. Similarly if one is interested in rotational invariance, the filter itself can be rotated, providing that capability.

It is often desirable to detect objects on the basis of edges rather than energies, the classic example being the detection of edge objects in the same aerial photograph with varying degrees of contrast. Such a situation naturally arises with bright illumination and the illumination obtained from overcast weather [9], and detection of objects on an edge criteria can be described by the gradient matched filter [10]. A class of matched filters, of which the gradient and traditional energy matched filters (derived above) are subsets, can be described by the transfer function

$$H_p(u, v) = (-1)^p \frac{(u^2 + v^2)^p F_I^*(u, v) \exp\{-j(u\xi + v\eta)\}}{S_{N_I}(u, v)} \qquad (6)$$

The generalized matched filter of Eq. (6) reduces to:

(1) the traditional energy matched filter for $p = 0$;
(2) the gradient matched filter for $p = 1$;
(3) the Laplacian matched filter for $p = 2$;
(4) the pth-order gradient matched filter for general p.

In terms of the energy matched filters, Eqs. (5a) or (5c), the above filters can be likened to a matched filter cascaded with a filter which is compensating for an optical system OTF whose response falls off as $(u^2 + v^2)^{-p}$. However, the higher order quadratic terms in the filter of Eq. (6) are also equivalent to differential operations in the spatial domain. Consequently the output of the filter, $\mathcal{O}_p(x, y)$, can be expressed as

$$\mathcal{O}_p(x, y) = [\nabla^p \{f^*(\xi - x, \eta - y)\}] \circledast [\nabla^p \{g(x, y)\}] \circledast Q_N(x, y) \qquad (7)$$

where ∇^p is the pth-order gradient operator.

$$\nabla^p = \frac{\partial^p}{\partial x^p} + \frac{\partial^p}{\partial y^p}$$

In a later section the gradient matched filter will be applied both as a detector and image evaluator in a digital environment for various test scenes.

4.2 IMPLEMENTATION

For optical data processing systems, implementation of a true matched filter can de extremely difficult. This is because the filter contains both modulus and phase information which somehow must be placed on some form of light sensitive material. Because of the difficulty in storing continuous phase information on such materials, optical matched filtering was never very successful until the advent of coherent light and the techniques of holography. Vander Lugt [6, 7] has shown how these two new phenomena can be utilized to implement a matched filter of the form of Eqs. (5a) or (5c). His approach is to use conventional amplitude and phase modulation techniques to store continuous modulus and phase filter information on a spatial rather than temporal carrier wave. By beating a reference wave of coherent light with the desired matched filter, $H(u, v)$, on a photosensitive film, the intensity pattern on the film becomes

$$I(u, v) = |R(u, v) + H(u, v)|^2 \qquad (8a)$$

When the reference is a unit amplitude plane wave the intensity becomes

$$I(u, v) = |\exp\{+j(bu + cv)\} + H(u, v)|^2 \qquad (8b)$$

or

$$I(u, v) = 1 + |H(u, v)|^2 + H^*(u, v)\exp\{-j(bu + cv)\} \\ + H(u, v)\exp\{j(bu + cv)\} \qquad (8c)$$

Notice that the last term of Eq. (8c) becomes the desired matched filter multiplied by a linear phase factor proportional to the two-dimensional spatial carrier determined by parameters b and c. By optically multiplying the above filter with the Fourier transform, $G(u, v)$, of the input signal, $g(x, y)$, and inverse transforming, three significant terms will emerge. Figure 1(a) is the conventional coherent optical filtering system. The output $\mathcal{O}(x, y)$ becomes

$$\mathcal{O}(x, y) = \mathcal{F}^{-1}\{G(u, v)[1 + |H(u, v)|^2]\} \\ + \mathcal{F}^{-1}\{G(u, v)H^*(u, v)\exp\{-j(bu + cv)\}\} \\ + \mathcal{F}^{-1}\{G(u, v)H(u, v)\exp\{+j(bu + cv)\}\} \qquad (9a)$$

The three terms in the above expression result in being spatially separated by the carrier frequencies b and c. The third term in Eq. (9a) is the desired filtered function, $\mathcal{O}_1(x, y)$.

$$\mathcal{O}_1(x, y) = g(x, y) \circledast h(x, y) \circledast \delta(x + b, y + c) \qquad (9b)$$

The experimental results of the above method of matched filter implementation are encouraging although optical noise, film nonlinearities, and positioning problems will often plague such a system.

One technique to avoid some of the optical difficulties in constructing (but not implementing) the matched filters is to use digital computers. The matched filter is mathematically calculated in the computer and then some rendition of the filter is placed on a two-dimensional output device for conversion to optical format. Because the matched filter is proportional to the two-dimensional Fourier transform of the signal to be detected $F(u, v)$, see Eqs. (5a), (5c), or (6), such a transform can

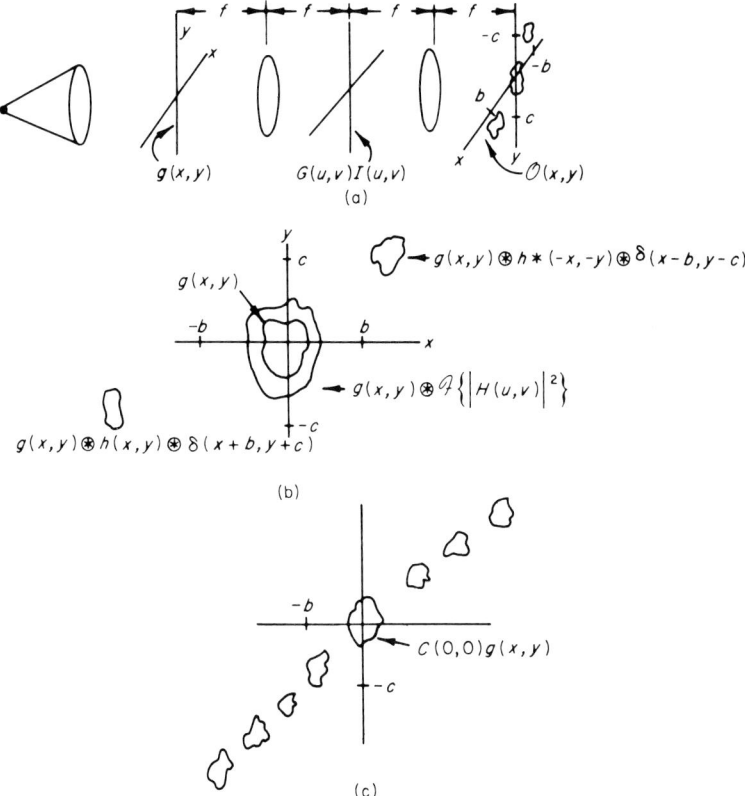

Fig. 1. Spatial carrier filters. (a) Optical filtering system. (b) Filter output plane. (c) Hard limited filter.

be computed, by the computer retaining all modulus and phase information necessary. The problem still exists as to how to store the information in an intensity format for photographic recording and further optical processing. Again the spatial carrier technique can be employed to store both phase and amplitude information raised on an average background intensity to insure no negative signals.

$$I(u, v) = S_b + H(u, v) \exp\{j(bu + cv)\} \\ + H^*(u, v) \exp\{-j(bu + cv)\} \quad (10)$$

where S_b is a bias term to insure nonnegativity and the $|H(u, v)|^2$ term of Eq. (8c) has been mathematically eliminated. If such a signal is now used in an optical system similar to that of Fig. 1, one of the central images, $g(x, y) \circledast \mathscr{F}\{|H(u, v)|^2\}$, will be missing, allowing lower spatial frequency carriers in the computer generated filter. The problem now exists as to how to get the filter of Eq. (10) out of the computer onto an optical format. The amplitude modulation requires a continuous tone rendition of the filter which is often difficult to develop. Consequently hard-limiting techniques can be employed such that the output is a binary signal retaining only phase modulation information where $H(u, v) = |H(u, v)| \exp\{j\phi(u, v)\}$.

$$I_L(u, v) = \frac{1}{2} + \frac{1}{2} \frac{|H(u, v)| \cos[bu + cv + \phi(u, v)]}{|H(u, v)| |\cos[bu + cv + \phi(u, v)]|} \quad (11a)$$

However, from nonlinear filtering theory [11] the hard-limiter above can be described as a full-wave odd νth law device. Such a filter introduces all odd order harmonics of the input and can be represented as

$$I_L(u, v) = \tfrac{1}{2} + \sum_{m \,(\text{odd})}^{\infty} C(0, m) \cos\{m[bu + cv + \phi(u, v)]\} \quad (11b)$$

where $\phi(u, v)$ is the phase information of the matched filter and where $C(0, m)$ is a constant as a function of m described by Davenport [11]. By using the Neumann factor, $\mathscr{E}_m = 2 - \delta(m)$ for all integer m, and representing the cosine in terms of exponentials, the filter becomes

$$I_L(u, v) = \sum_{m=0, \pm 1, \pm 3, \ldots} \frac{C(0, |m|)}{\mathscr{E}_m} \exp\{jm[bu + cv + \phi(u, v)]\} \quad (11c)$$

When this filter, $I_L(u, v)$, is optically multiplied by $G(u, v)$ and inverse transformed as in Fig. 1, the output will be an infinite number of two sided diffraction patterns separated by the spatial carriers determined by mb and mc.

$$\mathcal{O}(x, y) = \sum_{m=0, \pm 1, \pm 3, \ldots}^{\infty} \frac{C(0, |m|)}{\mathscr{E}_m} g(x, y) \circledast \mathscr{F}^{-1}\{\exp\{jm\phi(u, v)\}\}$$
$$\circledast \, \delta(x + mb, y + mc) \quad (11d)$$

The zero th-order output, $m = 0$, will consist of $C(0, 0)g(x, y)$ and is of no interest. The positive first-order term, $m = +1$, will consist of a matched filtered version of the input, $g(x, y)$, but matched only to the phase spectrum of the signal, $f(x, y)$, and independent of the amplitude spectrum.

$$\mathcal{O}_{+1}(x, y) = \tfrac{1}{2}C(0, 1)[g(x, y) \circledast \delta(x - b, y - c)$$
$$\circledast \, \mathscr{F}^{-1}\{\exp\{j\phi(u, v)\}\}] \quad (12)$$

While such a filter is not truly a matched filter, one-dimensional experimental implementation of these principles has resulted in surprisingly successful results [12]. The higher order terms for $m = \pm 3, \pm 5, \ldots$, are of little interest as they provide filtering with odd multiples of the phase function, $m\phi(u, v)$, spaced at odd multiple centers, $\delta(x + mb, y + mc)$, away from the origin and zeroth-order terms; see Fig. 1.

Another technique for using binary planes of information in spatial filtering operations has been developed by Lohmann [13-17]. His approach is to develop the binary filter plane from a two-dimensional diffraction grating analysis. A digital computer is used to calculate the positions and size of the slits in the grating to perform a "detour phase" effect upon optical illumination [15]. The computer plots out the required grating which is then optically reduced for optical bench illumination. The experimental results obtained with this technique are quite successful despite the traditional optical bench signal-to-noise ratio and illumination problems.

The last technique for implementation of the two-dimensional matched filtering process to be described is an all digital system. Because the computer has a tremendous signal-to-noise advantage over physical systems and because there is no need to implement the filtering process on energy sensitive films, the matched filter can simply be constructed with a two-dimensional Fourier transformation and conjugate operation of the signal to be detected. While computing such transformations requires quite powerful digital facilities, description of implementation of the necessary algorithms will be deferred until a later chapter.

Assuming that a two-dimensional Fourier transform system exists, construction of the filters of Eqs. (5a), (5c), or (6) is straightforward. In fact the effect of gradient operators, as discussed earlier, is accomplished by multiplying the energy matched filter of Eqs. (5a) or (5c) by a polynomial surface in the Fourier transform domain. The actual filtering operation, then, is obtained by Fourier transforming the input plane, multiplying by the filter plane, and then inverse transforming the product.

4.3 APPLICATIONS

The application of matched filters to one-dimensional signal processing has been alluded to earlier. In higher dimensions, the principles remain the same and conceivably the n-dimensional matched filter is a perfectly legitimate technique for application in pattern recognition. In the two-dimensional area, the matched filter has been described as useful for character recognition and image detection based on the shape and height of the signal to be distinguished. For a multiple image detection problem, parallel application of multiple matched filters results in the optimum detection scheme.

As mentioned earlier, it is often desirable to detect objects on the basis of edge or gradient information as well as enclosed or total energy information. Such an approach resulted in the modified matched filter of Eq. (6), repeated here for future reference.

$$H_p(u, v) = (-1)^p \frac{(u^2 + v^2)^p F_I^*(u, v)}{S_{N_I}(u, v)} \exp\{-j(u\xi + v\eta)\} \qquad (6)$$

This generalized filter has been simulated (assuming white noise, $S_{N_I}(u, v) = N_0/2$) on a digital computer for the two cases: $p = 0$, a traditional energy matched filter; and $p = 1$, a gradient matched filter.

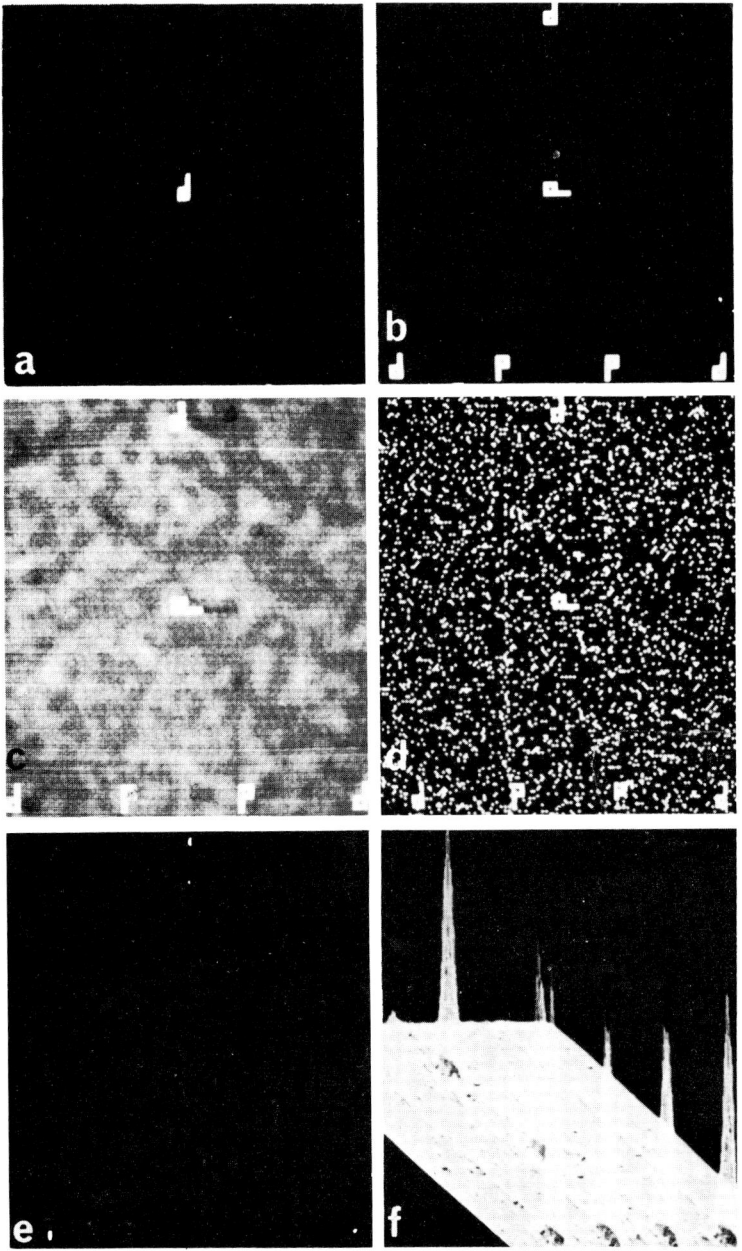

Fig. 2. Image detection. (a) Signal. (b) Array. (c) Array and low frequency noise. (d) Array and high frequency noise. (e) Filter output after threshold detection. (f) Perspective of correlation peaks.

The signal to be detected is the letter "d" displayed in Fig. 2(a). The purpose of the matched filtering is to detect all the "d's" out of the unknown signals presented to the filter. Three different unknowns have been filtered, each with the energy and gradient matched filters. The unknowns consist of an array of "d's" and "p's" on three different backgrounds as displayed in Figs. 2(b), 2(c), and 2(d). In all cases both the energy and gradient matched filters properly detected the three "d's" in the array after simple threshold detection on the filter output as shown in Fig. 2(e). A perspective of the filter output plane is shown in Fig. 2(f) in which the three largest peaks correspond to the proper matches in the original array. The ratios of signal peak to noise (rotated "d") peak for each unknown and each filter is given in Table I.

TABLE I

	Array b	Array c	Array d
Energy matched filter	1.330	1.319	1.260
Gradient matched filter	1.464	1.450	1.291

It is evident from the table that the gradient matched filter gives a stronger signal match than the energy matched filter for all three different background conditions. This can be explained by the fact that the signal is described by many edges and thus lends itself more readily to edge detection than traditional energy detection. It should be pointed out that the matched filters used in this experiment were designed for the white noise case. If the filters had been normalized by the proper spectrum of the noise in each of the above three test arrays, the signal matches would have been stronger.

One application for two-dimensional matched filters not already mentioned is that of image evaluation. For automatic processing and decision making concerning images, a quantitative measure of image evaluation must be made available. In the area of space exploration, photometry, fingerprint recognition, biomedical image processing, photo-reconnaissance, terrain mapping, and a variety of other applications, a measure of "goodness" must be made available for complete and fully automatic processing of image data. It is the author's opinion that the common engineering tool of mean square error evaluation is not practical for image processing especially if the human is to be the ultimate observer. It is a safe assumption that the eye does not compose images on a mean square difference criteria. It is suggested that some linear combination of gradient operations might be more meaningful if they could be lumped into one parameter. Such a criteria might result in the operation

$$\sum_{n=0}^{N} a_n \nabla^n \{f(x, y)\}$$

which has a convenient polynomial counterpart in the Fourier transform domain. Coupled with the concepts of matched filtering, the question of how good an image is with respect to a pre- or post-processed version need not be answered by as many subjectively different opinions as there are viewers. A two-dimensional correlation between images can be made with a generalized matched filter and the resulting correlation value will be a quantitative measure of the "distance" betwee the two images. Just as the positions of the correlation peaks in Fig. 2(f) indicate the degree of correlation of the detected signal with the pattern to be recognized, so then this latter property, the degree of correlation, can be used as a tool for automatic image evaluation. The output of the filter is given by Eq. (7) and for spatial registration of the two images, $\xi = \eta = 0$, the greatest correlation will occur at the origin, $x = y = 0$.

$$\mathcal{O}_p(0, 0) = \iint_{-\infty}^{\infty} \frac{(u^2 + v^2)^p F_I^*(u, v)}{S_{N_I}(u, v)} G(u, v) \, du \, dv \qquad (13)$$

The peak correlation will occur when the filter matches an image with itself in the absence of noise.

$$\mathcal{O}_p(0, 0)_{\max} = \iint_{-\infty}^{\infty} |F(u, v)|^2 (u^2 + v^2)^p \, du \, dv \qquad (14)$$

The ratio $\mathcal{O}_p(0, 0)/\mathcal{O}_p(0, 0)_{\max}$ can then be interpreted as a correlation parameter, and image evaluation decisions can be based on this ratio.

An example of the use of such correlation ratios is presented in Fig. 3. Both energy and gradient matched filter correlation ratios have

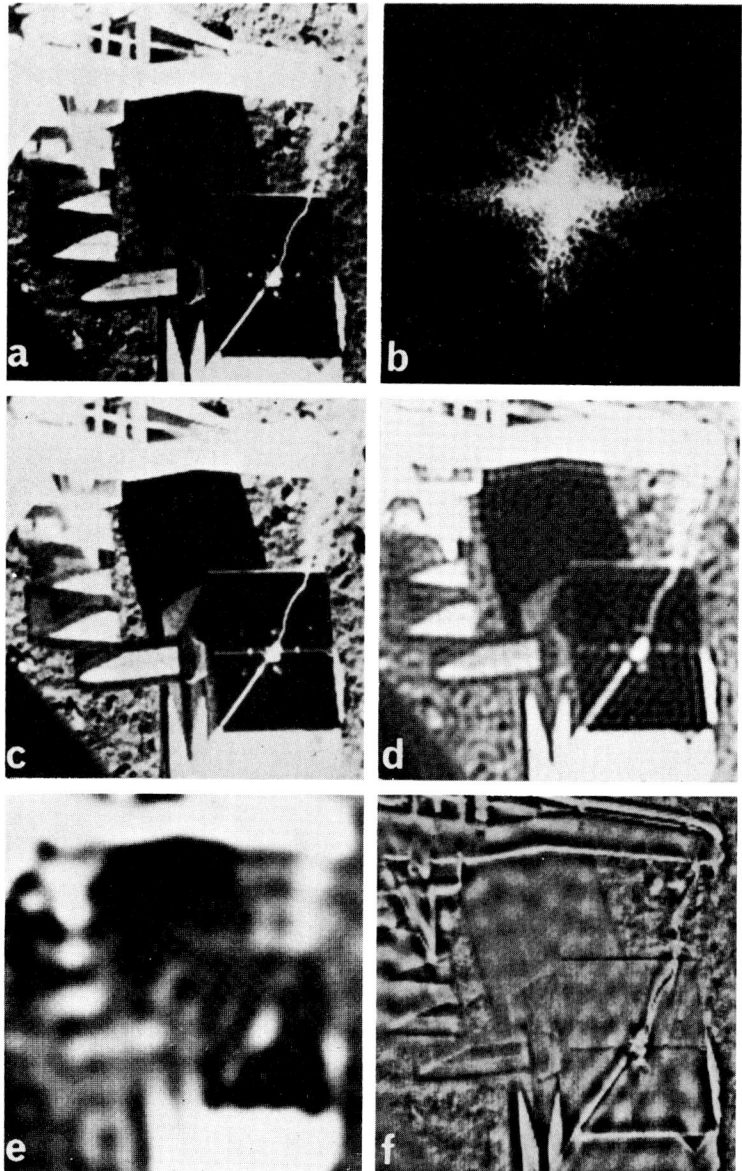

Fig. 3. Image evaluation (a) Original. (b) Logarithm of the magnitude of the Fourier transform. (c) Double Fourier transform energy correlation: 1,0000, gradient correlation: 1.0000. (d) Low pass 8:1 bandwidth reduction energy correlation: 0.9633; gradient correlation: 0.2633. (e) Low pass 64:1 bandwidth reduction energy correlation: 0.8659; gradient correlation: 0.0449. (f) High pass 1.6% bandwidth reduction energy correlation: 0.0611; gradient correlation: 0.2480.

been determined for four different test cases. In Fig. 3c the unknown scene matches perfectly with the original image and both energy and gradient ratios reflect this fact. The original image has been lowpass filtered in Fig. 3(d) and the correlation ratios indicate the effect of such a filtering operation. As would be expected, the gradient correlation drops substantially with the removal of the higher frequency data, while the large amplitude low frequency data remaining provide a good match for the energy matched filter correlation. The same phenomena is evident in Fig. 3(e) where even more high frequency data has been removed. As would be expected, the removal of low frequency high amplitude data reverses the trend. Figure 3(f) displays a high pass filtered version of the original with the lowest 1.6% of the spatial frequencies removed. The energy correlation ratio drops drastically while the gradient ratio indicates a much stronger comparative match.

The above discussion suggests a technique for computer application to image detection (interpretation) and evaluation using the Fourier transformation operation. The experimental results are encouraging but do not preclude other evaluation procedures. One's natural familiarity with spatial frequencies should not restrict the search for other mathematical transforms thal might be as applicable to automatic evaluation procedures. For instance, any set of two-dimensional orthonormal functions might serve as a domain for evaluation and photo-interpretation.

4.4 CONCLUSIONS

The concepts of matched filtering have been presented in this chapter in a two-dimensional environment. The filter has been derived for a maximization of signal to noise ratio and has been shown to be a detection device for all possible translations of a given input signal. The traditional matched filter bases its detection capabilities on an integrated energy criteria. When it becomes desirable to detect objects on the basis of edges rather than integrated energies, then the matched filter can be generalized to take such requirements into consideration. A generalized gradient matched filter is suggested for such operations.

Implementation of the matched filter can be developed in one of three general techniques. The first is a purely optical technique in which the holographic properties for storing phase and amplitude information are utilized for photosensitive film. A second technique which avoids some of the construction problems in the holographic system is through the use of a digital computer in computing the matched filter desired. The computed result is then output onto photographic film for optical filtering techniques. The most appealing matched filtering implementation from a mathematical and signal to noise basis is the use of purely digital computer techniques in which the matched filter is constructed in the computer and the filtering operation is also carried out within the computer. In such a case all degrees of freedom are utilized and very accurate filtering results. Examples of the application of matched filters in a purely digital environment are presented for both the energy and gradient filters described earlier. The filter is successfully used to detect translated objects in the presence of rotated objects and different noise spectra. Finally a brief descussion is developed concerning the automatic image evaluation through the use of matched filters. Again experimental results are implemented in order to verify the concepts set forth. It was suggested that possibly other techniques, unfamiliar to Fourier procedures, might be applicable to the automatic image evaluation task. Such techniques could use other two-dimensional orthonormal basis functions rather than the trigonometric functions as decomposition waveforms for image evaluation. The following chapter presents a variety of rapidly and efficiently implementable digital orthogonal transformations for possible application in the area of image evaluation and other forms of image processing.

REFERENCES

1. G. L. Turin, "An Introduction to Matched Filters," *IRE Transactions on Inform. Theory*, 311-329 (June 1960).
2. W. B. Davenport, Jr., and W. L. Root, *An Introduction to the Theory of Random Signals and Noise*, p 224. McGraw-Hill, New York, 1958.
3. C. E. Cook and M. Bernfeld, *Radar Signals*. Academic Press, New York, 1967.
4. J. M. Wozencraft and I. M. Jacobs, *Principles of Communication Engineering*, pp 234-244. Wiley, New York, 1965.

5. C. E. Cook and M. Bernfeld, *Radar Signals*, Chapter 2. Academic Press, New York, 1965.
6. A. Vander Lugt, F. B. Rotz, and A. Klooster, Jr. "Character-Reading by Optical Spatial Filtering," In *Optical Electro-Optical Information Processing*. (J. Tippitt et al. eds.), pp 125-141 MIT Press, Cambridge, Massachusetts, 1965.
7. A. Vander Lugt, "Signal Detection by Complex Spatial Filtering," *IEEE Trans. Inform. Theory* **IT-10,** No. 2, 139-145 (April 1964).
8. W. M. Brown, *Analysis of Linear Time-Invariant System*. McGraw-Hill, New York, 1963.
9. A. W. Lohmann and D. P. Paris, "Computer Generated Spatial Filters for Coherent Optical Data Processing," *Appl. Opt.* **7,** 651 (April 1968).
10. A. Marechal and P. Croce, *Compt. Rend. Acad. Sci.* **273,** 607 (1953).
11. W. B. Davenport, Jr., and W. L. Root, *An Introduction to the Theory of Random Signals and Noise*, Chapter 13. McGraw-Hill, New York, 1958.
12. A. Kozma and D. L. Kelly, "Spatial Filtering for Detection of Signals Submerged in Noise," *Appl. Opt.* **4,** No. 4, 387-392 (April 1965).
13. M. De and A. W. Lohmann, "Signal Detection by Correlation of Fresnel Diffration Patterns," *Appl. Opt.* **6,** No. 12, 2171-2175 (December 1967).
14. A. W. Lohmann, D. P. Paris, and H. W. Werlich, "A Computer Generated Spatial Filter Applied to Code Translation," *Appl. Opt.* **6,** No. 6, 1138-1140 (June 1967).
15. B. R. Brown and A. W. Lohmann, "Complex Spatial Filtering with Binary Masks," *Appl. Opt.* **5,** No. 6, 967-969, (June 1966).
16. A. W. Lohmann and D. P. Paris, "Computer Generated Spatial Filters for Coherent Optical Data Processing," *Appl. Opt.* **7,** No. 4, 651-655 (April 1968).
17. A. W. Lohmann, "Matched Filtering with Self-Luminous Objects," *Appl. Opt.* **7,** No. 3, 561-563 (March 1968).

Chapter 5 ORTHOGONAL TRANSFORMATIONS[1]

5.0 INTRODUCTION

Before the development of digital computers, few systems took real advantage of the power of the theory of orthogonal transformations other than traditional Fourier spectrum analyzers. In engineering applications, Fourier analysis has found major applications particularly in the study of time series. With the advent of digital processors, discrete Fourier decompositions were attempted but mathematical implementation proved time consuming and cumbersome. Recently, efficient algorithms have been developed which vastly decrease the computational requirements of Fourier analysis and renewed interest has been generated in digital spectral decomposition [1-6]. In addition, methods other than Fourier spectral analysis have begun to become prevalent in certain aspects of digital processing and the theory of orthogonal transformations is becoming an active area in which computer users are searching for powerful processing tools.

It is useful to study a few of the properties of orthogonal decompositions within a digital environment, and matrix theory offers a tool for such purposes. In the study of finite dimensional vector spaces, orthogonal transformations provide a preservation of inner products when interpreted as linear operators from one space to another. The inverse transformation is particularly simple to imple-

---

[1] This chapter is by Harry Andrews and Kenneth Caspari.

ment, being the matrix transpose rather than inverse. In addition the transformations are entropy preserving in an information theoretic sense, and when applied to certain pattern recognition systems, to be described later, the transformations provide rotations of the pattern space in which feature selection and maximum variance terchniques can be applied. A distinct advantage of orthogonal transformations, other than eliminating the need for matrix inversion, is that the coefficients of the orthogonal vectors defining the transformation matrix are independent and consequently, higher order approximations can be obtained without recalculating lower order coefficients. In generalized spectral analysis an orthogonal transformation provides a set of coefficients which indicate the degree of correlation of data vectors with each vector defining the transformation matrix. In digital communication and coding theory, orthogonal vectors themselves have become important candidates for signal design and error correcting codes. However, possibly one of the most appealing justifications for the use of orthogonal transformations lies in the resulting analysis of the eigenvalues and eigenvectors which define optimum solutions to various systems in the theory of stochastic approximations as well as conventional systems theory. In fact, we have met three such cases in the chapter on Fourier optics. Consider a linear time (shift) invariant system with impulse response $h(t)$. When the input function is of the form $e^{j\omega t}$, the resulting convolutional output becomes

$$g(t) = e^{j\omega t} H(\omega) \tag{1a}$$

where $H(\omega)$ are the eigenvalues and $e^{j\omega t}$ the eigenvectors of the linear system [7]. Notice that the set of eigenvectors form an orthogonal class of functions. The two-dimensional circularly symmetric linear system was seen to have an eigenvector solution when the input was the zero-order Bessel function of the first kind, $J_0(ar)$ [8]. Thus

$$g(r) = 2\pi \bar{h}(a) J_0(ar) \tag{1b}$$

where $\bar{h}(a)$, the Hankel transform of the impulse response of the system, is the eigenvalue associated with the eigenvector $J_0(ar)$ and the set of eigenvectors again form a complete orthogonal class. In the study of super resolution, it was found that the prolate spheroidal wavefunctions, $\psi_j(t)$, became the eigenvector solutions to the finite

Fourier transform process

$$\lambda_i \psi_i(t) = \int_{-T/2}^{T/2} \frac{\sin \omega(t-\tau)}{t-\tau} \psi_i(\tau) \, d\tau \qquad (1c)$$

where again the class of prolate spheroidal wavefunctions formed an orthogonal set [9]. Other examples of eigenvector solutions to specific systems are numerous. In the area of stochastic processes, the optimal solution to the bandwidth reduction task is given by the Karhunen-Loeve expansion theorem resulting in a set of eigenvalues and eigenvectors [10]. In discrete notation, given a sample covariance matrix $[R(s, t)]$, the eigenvector solution provides a set of orthogonal vectors with associated eigenvalues.

$$[R(s, t)]\bar{\phi} = \lambda \bar{\phi} \qquad (1d)$$

The matrix comprised of the eigenvector solution to the covariance matrix is orthogonal and becomes the optimal decomposition of signals from the class of signals described by the covariance matrix. The decomposition is optimum in a mean square error reconstructibility sense meaning that the truncation of the coefficients of the lowest eigenvalues, for a dimensionality or bandwidth reduction, reconstructs the best mean square approximation to the original signal.

While the eigenvector solution to specific systems or processes provides optimum orthogonal decomposition matrices, calculations of the eigenvalues and their associated vectors is a very difficult if not numerically impossible task. Consequently often suboptimum in a mean square error sense, decompositions are desirable if they can be efficiently and conveniently generated by digital processes. This then is the object of this chapter. A system of orthogonal transformations will be presented which are implementable in $pN \log_p N$ or less operations compared to the normal N^2 operations required for arbitrary linear transformations. The class of transformations so implementable include the Fourier, Hadamard, Walsh, generalized Walsh, Haar, generalized Haar, and a variety of unnamed classes of orthogonal transformations. Examples are presented to display some interesting "transition" orthogonal transformations and a brief quantum level analysis is developed to further describe certain of these transformations. A generalized spectrum analyzer is hypothesized after which a variety of applications are suggested. The specific area of principal component and maximum variance analysis for pattern recognition applications is described and a possible adaptive feature selection and search technique using the fast transform techniques is suggested.

5.1 FAST TRANSFORMATIONS[2]

The underlying principle for the efficient implementation of the transformations presented here is a high degree of redundancy in the transform matrix description. If the redundancy in the definition of the matrix transform can be eliminated by matrix factorization, then a more efficient means of implementation is available. Such a technique was described by Good [12] in 1958 and has resulted in the fast Fourier transform, FFT [1], and fast Hadamard transform, FHT [13], as well as a much larger class of fast transformations described here.

Decomposition of an input waveform into a set of coefficients of orthogonal waveforms for generalized spectral analysis is equivalent to a vector-matrix multiplication if implemented on a digital computer. If the matrix defining the decomposition can be factored into a kronecker product of a set of matrices then implementation of the transformation can be accomplished with a reduced number of arithmetic operations.

Consider the class of matrices formed by the kronecker product operation. Let the submatrices be square and of dimension p by p with entries $m_{r,i,j}$ where i and j range from 0 to $p - 1$.

$$M_r = \begin{bmatrix} m_{r,0,0} & m_{r,0,1} & \cdots & m_{r,0,p-1} \\ m_{r,1,0} & & & \\ \vdots & & & \\ m_{r,p-1,0} & \cdots & & m_{r,p-1,p-1} \end{bmatrix} \quad (2)$$

Here the first index represents the class of entries corresponding to

---

[2] From Andrews and Kane [11], "Kronecker Matrices, Computer Implementation, and Generalized Spectra," *J. Assoc. Computing Machinery* (April 1970). Copyright ©, 1970, Association for Computing Machinery, Inc., New York.

a particular dimension in the Kronecker product operation. In general,

$$H_1 = M_0 \tag{3a}$$
$$H_2 = M_1 \otimes H_1 \tag{3b}$$
$$\vdots$$
$$H_n = M_{n-1} \otimes H_{n-1} \tag{3c}$$

where $\otimes$ is the Kronecker product operator [14]. Thus

$$H_n = \begin{bmatrix} m_{n-1,0,0} H_{n-1} & \cdots & m_{n-1,0,p-1} H_{n-1} \\ m_{n-1,1,0} H_{n-1} & \cdots & m_{n-1,1,p-1} H_{n-1} \\ \vdots & & \vdots \\ m_{n-1,p-1,0} H_{n-1} & \cdots & m_{n-1,p-1,p-1} H_{n-1} \end{bmatrix} \tag{4}$$

where H_n is a p^n by p^n matrix.

When operating with Kronecker matrices within a computer, it becomes desirable to store a representation (algorithm) of the entries of the product matrix rather than the matrix itself. Towards this end, consider the locations in the matrix to be described by their lexicographic or dictionary sequence representation. In other words, a given index of matrix H_n can be represented by n digits each of which can take on the value 0 to $p - 1$. Representing the horizontal index by u and the vertical index by x, the names of the rows and columns in dictionary sequence for the H_2 matrix with $p = 3$ are

$$H_2 = \begin{array}{c} \\ \\ x \\ \\ \\ \\ \\ \\ \end{array} \begin{array}{c} 00 \\ 01 \\ 02 \\ 10 \\ 11 \\ 12 \\ 20 \\ 21 \\ 22 \end{array} \begin{array}{c} u \longrightarrow \\ \begin{array}{ccccccccc} 00 & 01 & 02 & 10 & 11 & 12 & 20 & 21 & 22 \end{array} \\ \begin{bmatrix} & & & & & & & & \\ & & & & & & & & \\ & & & & H_2(x,u) & & & & \\ & & & & & & & & \\ & & & & & & & & \end{bmatrix} \end{array} \tag{5}$$

Representing the u and x variables in the dictionary number system, mod p requires n digits to allow u and x to range over 0 to p^n. Therefore u and x can be described by

$$u = u_{n-1}\ u_{n-2}\cdots u_1 u_0, \qquad u_i \varepsilon \{0, 1, \ldots, p-1\} \qquad (6a)$$

$$x = x_{n-1}\ x_{n-2}\cdots x_1 x_0, \qquad x_i \varepsilon \{0, 1, \ldots, p-1\} \qquad (6b)$$

Using such a notation allows the entries of the p by p matrix H_1 [Eq. (3a)] to be described by the equation

$$H_1(x, u) = \prod_{i=0}^{p-1} \prod_{j=0}^{p-1} m_{0,i,j}^{\delta(x_0 - i)\,\delta(u_0 - j)} \qquad (7)$$

where $\delta(a - b)$ is the delta function which takes on the value 1 whenever $a = b$ and 0 otherwise. The representation of Eq. (7) can be interpreted as multiplying all entries of the core matrix, Eq. (3a), together and noting that all but one entry will be raised to the 0th power. The entries of the p^2 by p^2 matrix, H_2 (Eq. (3b)), can now be represented as

$$H_2(x, u) = \prod_{i=0}^{p-1} \prod_{j=0}^{p-1} m_{1,i,j}^{\delta(x_1 - i)\,\delta(u_1 - j)} \prod_{i=0}^{p-1} \prod_{j=0}^{p-1} m_{0,i,j}^{\delta(x_0 - i)\,\delta(u_0 - j)} \qquad (8)$$

where again the exponents determine the correct product of entries for a given u and x. In general, the entries for H_n can be represented as

$$H_n(x, u) = \prod_{r=0}^{n-1} \prod_{i=0}^{p-1} \prod_{j=0}^{p-1} m_{r,i,j}^{\delta(x_r - i)\,\delta(u_r - j)} \qquad (9)$$

following the recursive notation of Eqs. (7) and (8). Representation of the rows or columns of a Knonecker matrix in the form of Eq. (9) now allows the generation of any single element, column, or row of the matrix without storage of the entire matrix array. This becomes particularly important for large matrices especially in the area of generalized spectral analysis. The operation indicated in Eq. (9) can be envisioned as multiplying all entries of the submatrices, M_r, forming the kronecker product, together and letting the exponent operation allow only the proper entries to be raised to the power 1 while all others are raised to the power 0. For the Kronecker product of identical matrices

$$H_n(x, u) = \prod_{i=0}^{p-1} \prod_{j=0}^{p-1} m_{i,j}^{\Sigma_{r=0}^{n-1} \delta(x_r - i) \delta(u_r - j)} \tag{10}$$

In order to require $H_n(x, u)$ to be orthogonal, it is sufficient that the set of matrices, M_r, satisfy the orthogonality requirement for all $r = 0, 1, \ldots, n - 1$.

For $p = 2$, the resulting Kronecker matrices have resolution $N = 2^n$ and the closed product form of Eq. (9) becomes particularly convenient to describe. Let the core matrix H_1 be

$$H_1 = \begin{bmatrix} A_0 & B_0 \\ C_0 & D_0 \end{bmatrix} \tag{11}$$

and H_n can be represented as

$$H_n = \begin{bmatrix} A_{n-1} H_{n-1} & B_{n-1} H_{n-1} \\ C_{n-1} H_{n-1} & D_{n-1} H_{n-1} \end{bmatrix} \tag{12}$$

The closed form product representation now becomes

$$H_n(x, u) = \prod_{r=0}^{n-1} A_r^{\bar{x}_r \bar{u}_r} B_r^{\bar{x}_r u_r} C_r^{x_r \bar{u}_r} D_r^{x_r u_r} \tag{13}$$

where the exponent operations become Booleam "and" operations, and the bar over the binary variable represents the complement value. For the case in which $A_r = A_s$, $B_r = B_s$, $C_r = C_s$, $D_r = D_s$ for all r and s as in Eq. (10), the representation again simplifies and becomes

$$H_n(x, u) = A^{\Sigma_{r=0}^{n-1} \bar{x}_r \bar{u}_r} B^{\Sigma_{r=0}^{n-1} \bar{x}_r u_r} C^{\Sigma_{r=0}^{n-1} x_r \bar{u}_r} D^{\Sigma_{r=0}^{n-1} x_r u_r} \tag{14}$$

The orthogonality constraint becomes

$$A^2 + B^2 = 1 \tag{15a}$$
$$C^2 + D^2 = 1 \tag{15b}$$
$$AC + BD = 0 \tag{15c}$$

If it is desirable to make the resulting matrix symmetric and orthogonal so that a transformation taken twice results in the identity operation, then further simplifications result in the closed form representation of the matrix H_n. The additional symmetry constraint requires

$$B = C \tag{15d}$$

$$A = -D \quad \text{or} \quad B = 0 \tag{15e}$$

For the $B = 0$ solution, the resulting $H_n(x, u)$ matrix become diagonal

$$H_n(x, u) = \left(\frac{D}{A}\right)^{\sum_{r=0}^{n-1} u_r} \delta(u - x) \tag{16}$$

for the $A = -D$ solution the orthogonal symmetric matrix, $H_n(x, u)$, becomes

$$H_n(x, u) = A^n \left(\frac{(1 - A^2)}{A^2}\right)^{\frac{1}{2}\sum_{r=0}^{n-1} u_r \oplus x_r} (-1)^{\sum_{r=0}^{n-1} u_r x_r} \tag{17}$$

where $\oplus$ implies an exclusive "or" Boolean operation. The class of orthogonal matrices described by Eq. (17) is a one-parameter family of sets of Kronecker matrices. Consequently, valid 2-tuples defining such matrices are $\{\cos \theta, \sin \theta\}$, $\{1, 0\}$, $\{3/5, 4/5\}$, $\{1/\sqrt{2}, 1/\sqrt{2}\}$ and many others.

One transformation of particular interest is that given by the 2-tuple $\{\cos \theta, \sin \theta\}$, which represents sub matrices of

$$M_r = \begin{bmatrix} \cos \theta & \sin \theta \\ \sin \theta & -\cos \theta \end{bmatrix} \tag{18a}$$

As θ varies from 0 to 45 degrees, the resulting Kronecker matrix varies from a diagonal matrix to one in which the energy in each row (and column) is uniformly distributed over every entry. The matrix is given by

$$H_n(x, u) = \cos^n \theta \left(\frac{\sin \theta}{\cos \theta}\right)^{\sum_{r=0}^{n-1} u_r \oplus x_r} (-1)^{\sum_{r=0}^{n-1} u_r x_r} \tag{18b}$$

When $\theta = 0$ degree, we use the fact that 0 raised to the 0th power is 1, and when $\theta = 45$ degrees, the matrix reduces to the Hadamard matrix of order 2 which is equivalent to the discrete Walsh transform

[15]. The matrix of Eq. (18b) then becomes

$$H_n(x, u) = \left(\frac{1}{2}\right)^{n/2} (-1)^{\sum_{r=0}^{n-1} u_r x_r} \tag{19a}$$

which is generated by the Kronecker product of

$$M_r = \frac{1}{\sqrt{2}} \begin{bmatrix} 1 & 1 \\ 1 & -1 \end{bmatrix} \tag{19b}$$

with itself n times. Two-dimensional implementation of Eq. (18b) has been used for image transformation on a digital computer in the search for optimum decomposition functions for bandwidth reduction [16, 17].

The transformation described by Eq. (19a) also describes a class of error correcting codes given by the Hadamard matrices of order 2. Another example of a Hadamard matrix which can be easily represented in the above described lexicographic notation is the powers of four matrix generated by Kronecker products of

$$M_r = \begin{bmatrix} 1 & 1 & 1 & -1 \\ 1 & 1 & -1 & 1 \\ 1 & -1 & 1 & 1 \\ -1 & 1 & 1 & 1 \end{bmatrix} \tag{20a}$$

for all r. In this case, Eq. (10) describes the matrix in closed product form where $p = 4$ and $m_{i,j} = -1$ for all $i + j = 3$ and $m_{i,j} = 1$ otherwise. Consequently

$$H_n(x, u) = (-1)^{\sum_{r=0}^{n-1} \delta(x_r + u_r - 3)} \tag{20b}$$

where x_r and u_r range from 0 to 3. This particular transformation has the property that each orthogonal vector in the matrix H_n has approximately the same number of zero crossings. This is to be contrasted with the Hadamard or Walsh transform which has $N = 2^n$ different number of zero crossings. The Walsh transform equation (19a) has been generalized to a much larger class of orthogonal transformations by Chrestenson [18] who has described many of the convergence properties of this expanded class. In discrete matrix notation the generalized Walsh transforms of order p can be generated by the core matrix

$$M_r = \begin{bmatrix} W^0 & W^0 & \cdots & W^0 \\ W^0 & W^1 & \cdots & W^{p-1} \\ W^0 & W^2 & \cdots & W^{2(p-1)} \\ \vdots & \vdots & & \vdots \\ W^0 & W^{p-1} & \cdots & W^{(p-1)^2} \end{bmatrix} = [W^{ux}] \quad (21a)$$

where $W = \exp\{2\pi j/p\}$ and simplifications can be made due to the fact that $W^{ux} = W^{ux \bmod p}$. For an N by N discrete generalized Walsh transform where $N = p^n$, the matrix is given by

$$H_n(u, x) = \exp\left[\left\{\frac{2\pi(-1)^{1/2}}{p}\right\} \sum_{i=0}^{p-1} \sum_{j=0}^{p-1} \sum_{r=0}^{n-1} ij\, \delta(x_r - i)\, \delta(u_r - j)\right] \quad (21b)$$

as is evident from Eq. (10). Note that the discrete generalized Walsh transform of order 2 reduces to the Hadamard transform. It is also interesting to note that the generalized Walsh transform core matrix of order p performs a Fourier transform of resolution p. However, the Kronecker product producing the generalized Walsh transform, Eq. (21b), no longer performs a Fourier transform.

A major premise suggested earlier was that if highly redundant matrices could be factored into a product of matrices with few non-zero entries, then a fewer number of operations would be necessary for transformation implementation. Good [12] has shown that for matrices which have a certain degree of redundancy and which have resolution equal to a highly composite number, they can be factored into a product of matrices which allow vector-matrix multiplication on the order of $pN \log_p N$ operations as compared to N^2 operations where $N = p^n$. Good's paper has laid the foundation for the fast Fourier transform and fast Hadamard or Walsh transform.

Good's technique of matrix factorization can be used to decompose the class of Kronecker matrices described by Eqs. (9) or (10). Thus, for the H_n Kronecker matrix of Eq. (9), there exist n matrices, each of dimension p^n, such that when multiplied together will equal H_n.

These matrices can be described as

$$G_r = \begin{bmatrix} m_{r,0,0} \cdots m_{r,0,p-1} & & & & \\ & m_{r,0,0} \cdots m_{r,0,p-1} & & & \\ & & \ddots & & \\ & & & m_{r,0,0} \cdots m_{r,0,p-1} & \\ m_{r,1,0} \cdots m_{r,1,p-1} & & & & \\ & m_{r,1,0} \cdots m_{r,1,p-1} & & & \\ & & \ddots & & \\ & & & m_{r,1,0} \cdots m_{r,1,p-1} & \\ \cdots & & & & \\ & \cdots & & & \\ & & \ddots & & \\ m_{r,p-1,0} \cdots m_{r,p-1,p-1} & & & & \\ & m_{r,p-1,0} \cdots m_{r,p-1,p-1} & & & \\ & & \ddots & & \\ & & & m_{r,p-1,0} \cdots m_{r,p-1,p-1} \end{bmatrix}$$

(22)

In this matrix there are p^{n+1} nonzero entries and only p^2 nonredundant elements. Then

$$H_n = [G_{n-1}][G_{n-2}] \cdots [G_1][G_0] \tag{23}$$

and for the Kronecker product of identical matrices

$$H_n = [G]^n \tag{24}$$

If a vector is multiplied by H_n, then N^2 operations will be required, whereas if the vector is multiplied by G_{n-1}, then pN operations will be required. If the resulting vector is multiplied by G_{n-2}, another pN operations will be required. If this step is carried out $n = \log_p N$ times, then a total of $pN \log_p N$ operations are necessary.

As an example, consider the discrete Walsh transform equivalent to the Hadamard matrix of dimensionality equal to a power 2, $N = 2^n$. Such matrices can be factored into one matrix, of only $2N$ nonzero

84 ORTHOGONAL TRANSFORMATIONS

entries, raised to the nth power. For $N = 8$, the Walsh–Hadamard matrix factors in the following way.

$$H_3 = \begin{bmatrix} 1 & 1 & 1 & 1 & 1 & 1 & 1 & 1 \\ 1 & -1 & 1 & -1 & 1 & -1 & 1 & -1 \\ 1 & 1 & -1 & -1 & 1 & 1 & -1 & -1 \\ 1 & -1 & -1 & 1 & 1 & -1 & -1 & 1 \\ 1 & 1 & 1 & 1 & -1 & -1 & -1 & -1 \\ 1 & -1 & 1 & -1 & -1 & 1 & -1 & 1 \\ 1 & 1 & -1 & -1 & -1 & -1 & 1 & 1 \\ 1 & -1 & -1 & 1 & -1 & 1 & 1 & -1 \end{bmatrix}$$

$$= \begin{bmatrix} 1 & 1 & 0 & 0 & 0 & 0 & 0 & 0 \\ 0 & 0 & 1 & 1 & 0 & 0 & 0 & 0 \\ 0 & 0 & 0 & 0 & 1 & 1 & 0 & 0 \\ 0 & 0 & 0 & 0 & 0 & 0 & 1 & 1 \\ 1 & -1 & 0 & 0 & 0 & 0 & 0 & 0 \\ 0 & 0 & 1 & -1 & 0 & 0 & 0 & 0 \\ 0 & 0 & 0 & 0 & 1 & -1 & 0 & 0 \\ 0 & 0 & 0 & 0 & 0 & 0 & 1 & -1 \end{bmatrix}^3 \quad (25)$$

Similarly, the generalized Walsh transform can be factored into simple matrices. Consider the generalized Walsh transform of order 3 where $N = 3^2$. The core matrix is given by

$$H_1 = \begin{bmatrix} W^0 & W^0 & W^0 \\ W^0 & W^1 & W^2 \\ W^0 & W^2 & W^1 \end{bmatrix} \quad (26)$$

and the transformation matrix and factored version becomes

$H_2 = H_1 \otimes H_1$

$$= \begin{bmatrix} W^0 & W^0 & W^0 & W^0 & W^0 & W^0 & W^0 & W^0 & W^0 \\ W^0 & W^1 & W^2 & W^0 & W^1 & W^2 & W^0 & W^1 & W^2 \\ W^0 & W^2 & W^1 & W^0 & W^2 & W^1 & W^0 & W^2 & W^1 \\ W^0 & W^0 & W^0 & W^1 & W^1 & W^1 & W^2 & W^2 & W^2 \\ W^0 & W^1 & W^2 & W^1 & W^2 & W^0 & W^2 & W^0 & W^1 \\ W^0 & W^2 & W^1 & W^1 & W^0 & W^2 & W^2 & W^1 & W^0 \\ W^0 & W^0 & W^0 & W^2 & W^2 & W^2 & W^1 & W^1 & W^1 \\ W^0 & W^1 & W^2 & W^2 & W^0 & W^1 & W^1 & W^2 & W^0 \\ W^0 & W^2 & W^1 & W^2 & W^1 & W^0 & W^1 & W^0 & W^2 \end{bmatrix}$$

$$= \begin{bmatrix} W^0 & W^0 & W^0 & 0 & 0 & 0 & 0 & 0 & 0 \\ 0 & 0 & 0 & W^0 & W^0 & W^0 & 0 & 0 & 0 \\ 0 & 0 & 0 & 0 & 0 & 0 & W^0 & W^0 & W^0 \\ W^0 & W^1 & W^2 & 0 & 0 & 0 & 0 & 0 & 0 \\ 0 & 0 & 0 & W^0 & W^1 & W^2 & 0 & 0 & 0 \\ 0 & 0 & 0 & 0 & 0 & 0 & W^0 & W^1 & W^2 \\ W^0 & W^2 & W^1 & 0 & 0 & 0 & 0 & 0 & 0 \\ 0 & 0 & 0 & W^0 & W^2 & W^1 & 0 & 0 & 0 \\ 0 & 0 & 0 & 0 & 0 & 0 & W^0 & W^2 & W^1 \end{bmatrix}^2 \quad (27)$$

Another class of transformations not already discussed but which have very efficient factorizations is known as the Haar transform [19]. The orthogonal but nonorthonormal Haar matrix is comprised of

86 ORTHOGONAL TRANSFORMATIONS

1's − 1's, and 0's and is directly related to the Walsh transform [20]. It is an orthogonal transformation, and in terms of sampling theory, samples the input waveform at progressively coarser intervals starting with the highest resolution and decreasing in powers of two. An example of the orthonormal Haar matrix for $N = 2^3$ in factored and final form is seen as

$$A_1 = \begin{bmatrix} 1 & 1 & 0 & 0 & 0 & 0 & 0 & 0 \\ 1 & -1 & 0 & 0 & 0 & 0 & 0 & 0 \\ 0 & 0 & \sqrt{2} & 0 & 0 & 0 & 0 & 0 \\ 0 & 0 & 0 & \sqrt{2} & 0 & 0 & 0 & 0 \\ 0 & 0 & 0 & 0 & \sqrt{2} & 0 & 0 & 0 \\ 0 & 0 & 0 & 0 & 0 & \sqrt{2} & 0 & 0 \\ 0 & 0 & 0 & 0 & 0 & 0 & \sqrt{2} & 0 \\ 0 & 0 & 0 & 0 & 0 & 0 & 0 & \sqrt{2} \end{bmatrix} \qquad (28a)$$

$$A_2 = \begin{bmatrix} 1 & 1 & 0 & 0 & 0 & 0 & 0 & 0 \\ 0 & 0 & 1 & 1 & 0 & 0 & 0 & 0 \\ 1 & -1 & 0 & 0 & 0 & 0 & 0 & 0 \\ 0 & 0 & 1 & -1 & 0 & 0 & 0 & 0 \\ 0 & 0 & 0 & 0 & \sqrt{2} & 0 & 0 & 0 \\ 0 & 0 & 0 & 0 & 0 & \sqrt{2} & 0 & 0 \\ 0 & 0 & 0 & 0 & 0 & 0 & \sqrt{2} & 0 \\ 0 & 0 & 0 & 0 & 0 & 0 & 0 & \sqrt{2} \end{bmatrix} \qquad (28b)$$

$$A_3 = \begin{bmatrix} 1 & 1 & 0 & 0 & 0 & 0 & 0 & 0 \\ 0 & 0 & 1 & 1 & 0 & 0 & 0 & 0 \\ 0 & 0 & 0 & 0 & 1 & 1 & 0 & 0 \\ 0 & 0 & 0 & 0 & 0 & 0 & 1 & 1 \\ 1 & -1 & 0 & 0 & 0 & 0 & 0 & 0 \\ 0 & 0 & 1 & -1 & 0 & 0 & 0 & 0 \\ 0 & 0 & 0 & 0 & 1 & -1 & 0 & 0 \\ 0 & 0 & 0 & 0 & 0 & 0 & 1 & -1 \end{bmatrix} \qquad (28c)$$

$H = [A_1][A_2][A_3]$

$$= \begin{bmatrix} 1 & 1 & 1 & 1 & 1 & 1 & 1 & 1 \\ 1 & 1 & 1 & 1 & -1 & -1 & -1 & -1 \\ \sqrt{2} & \sqrt{2} & -\sqrt{2} & -\sqrt{2} & 0 & 0 & 0 & 0 \\ 0 & 0 & 0 & 0 & \sqrt{2} & \sqrt{2} & -\sqrt{2} & -\sqrt{2} \\ 2 & -2 & 0 & 0 & 0 & 0 & 0 & 0 \\ 0 & 0 & 2 & -2 & 0 & 0 & 0 & 0 \\ 0 & 0 & 0 & 0 & 2 & -2 & 0 & 0 \\ 0 & 0 & 0 & 0 & 0 & 0 & 2 & -2 \end{bmatrix}$$

(28d)

The number of computer operations necessary to implement a Haar transformation is even less than $2N \log_2 N$ operations and is given by $2(N - 1)$ operations. For the example presented in Eqs. (28a), (28b), (28c), (28d) it is evident that the $[A_1]$ matrix requires 2 additions, the $[A_2]$ matrix requires 4 additions, and the $[A_3]$ matrix requires 8 additions resulting in a total of $2(8 - 1) = 14$ operations. As with the Walsh functions, the Haar functions can be generalized to contain entries of roots of unity other than ± 1. Watari [21] has described the generalized Haar system and has shown it is possible to preserve some of the original Haar convergence properties. The extension to matrix factorization is straightforward and will not be pursued further. However, the number of operations necessary to implement a pth-order generalized Haar transform is given by a geometric progression resulting in $p(p^n - 1)/(p - 1)$.

The last but very important example of matrix factorization to be described is that factorization technique which results in the fast Fourier transform. While it is not the intent of this chapter to delve into a complete description of a topic already adequately presented in the literature [1-6], it is interesting to note the parallelism of the fast Fourier technique and the method described here. In fact, with a minor modification in implementation, the core matrix necessary for the Walsh transform is identical to that which generates the Fourier transform of resolution equal to a power of 2. The modification necessary to implement the Fourier factorization in analogy with Eq. (23) for the Kronecker transform factorization is simply to introduce a scalar (or diagonal matrix) multiplication at each stage of the Good

ORTHOGONAL TRANSFORMATIONS

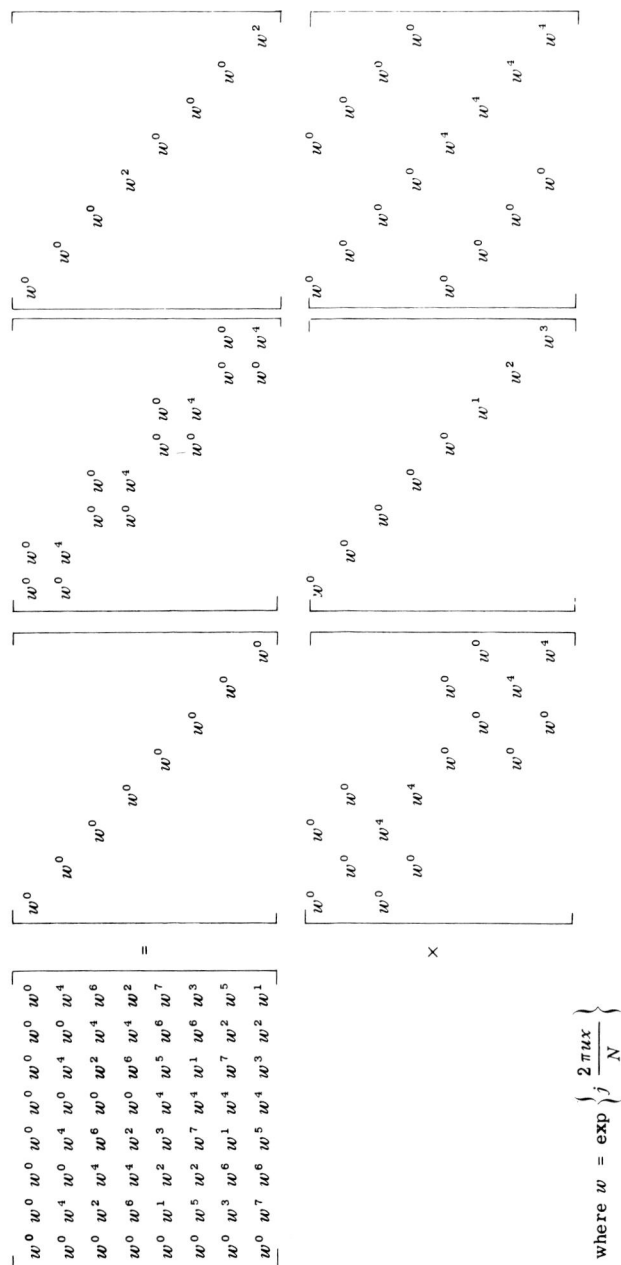

Fig. 1. Factorization of the 8 by 8 Fourier transform matrix

5.1 FAST TRANSFORMATIONS 89

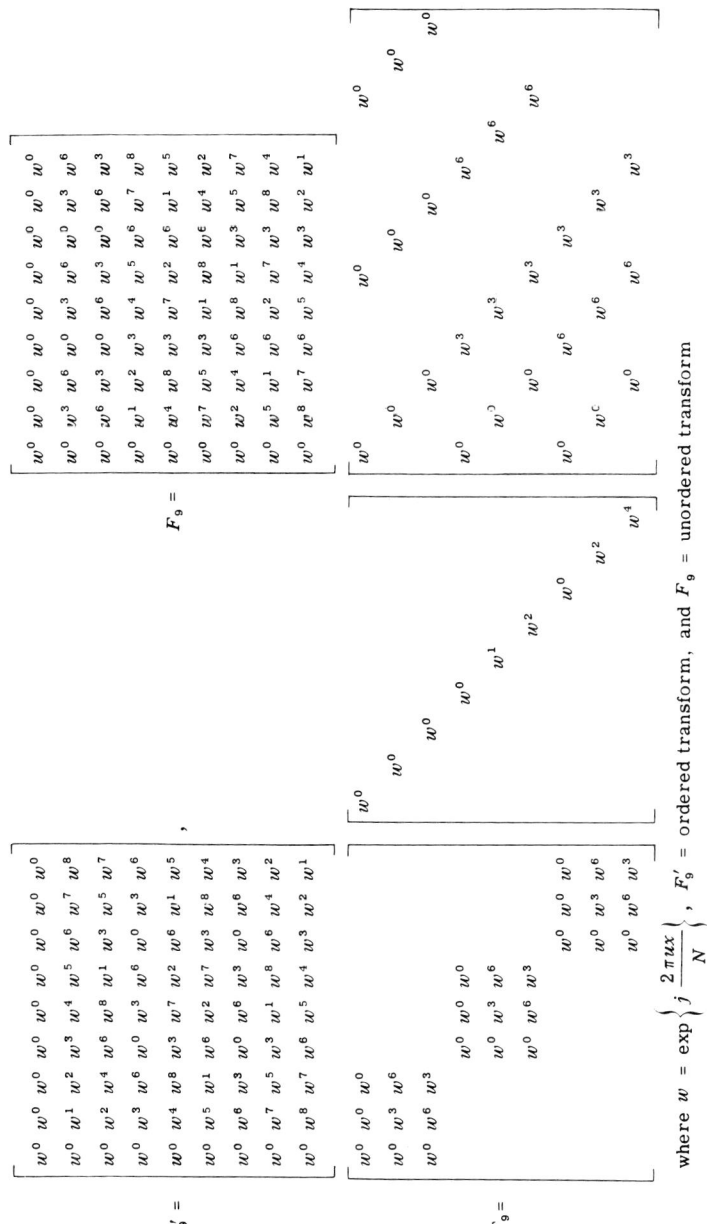

Fig. 2. Factorization of the 9 by 9 Fourier transform matrix. F_9' = ordered transform, and F_9 = unordered transform where $w = \exp\left\{j\dfrac{2\pi ux}{N}\right\}$.

decomposition [13]. Thus the Fourier transform matrix is given by

$$F_N = [I][G_0][D_1][G_1] \cdots [D_{n-1}][G_{n-1}] \qquad (29)$$

where $N = p^n$ and the Good matrices $[G_i]$ as well as the diagonals $[D_1]$ are deterministic permutations of each other, respectively. These permutations are described by Whelchel[13]. Examples of factorization of the Fourier matrix for $N = 2^3$ and $N = 3^2$ are presented in Figs. 1 and 2.

5.2 GENERALIZED SPECTRAL ANALYSIS[3]

The results of the Kronecker matrix and factorization sections suggest an efficient implementation of a whole class of transformations whose redundancies can be described in terms of either Kronecker products or Good matrices. Although such transformations need not be orthogonal, those that do have the orthogonality property can be used to form a generalized class of spectrum analyzers, examples of such include those cited earlier and of course many more. Implementation of such a general spectrum analyzer is hypothesized and a black box approach at this stage will be taken. Consider the block diagram in Fig. 3. Inputs to the analyzer are a p by p core matrix, a reference

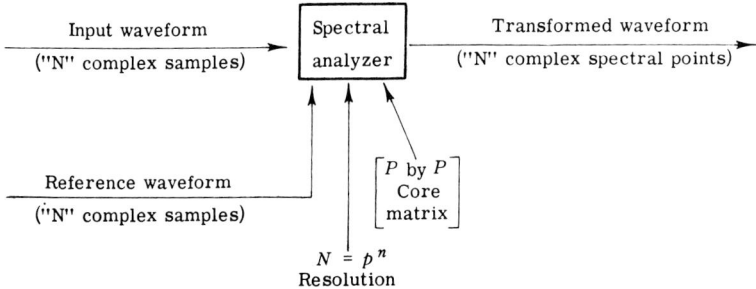

Fig. 3. General spectral analyzer and synthesizer.

---

[3] From Andrews and Caspari [22], "A Generalized Technique for Spectral Analysis," *IEEE Trans. Computers* **C-19**, No. 1, 16–25 (Jan. 1970). (Reprinted by permission.)

5.2 GENERALIZED SPECTRAL ANALYSIS **91**

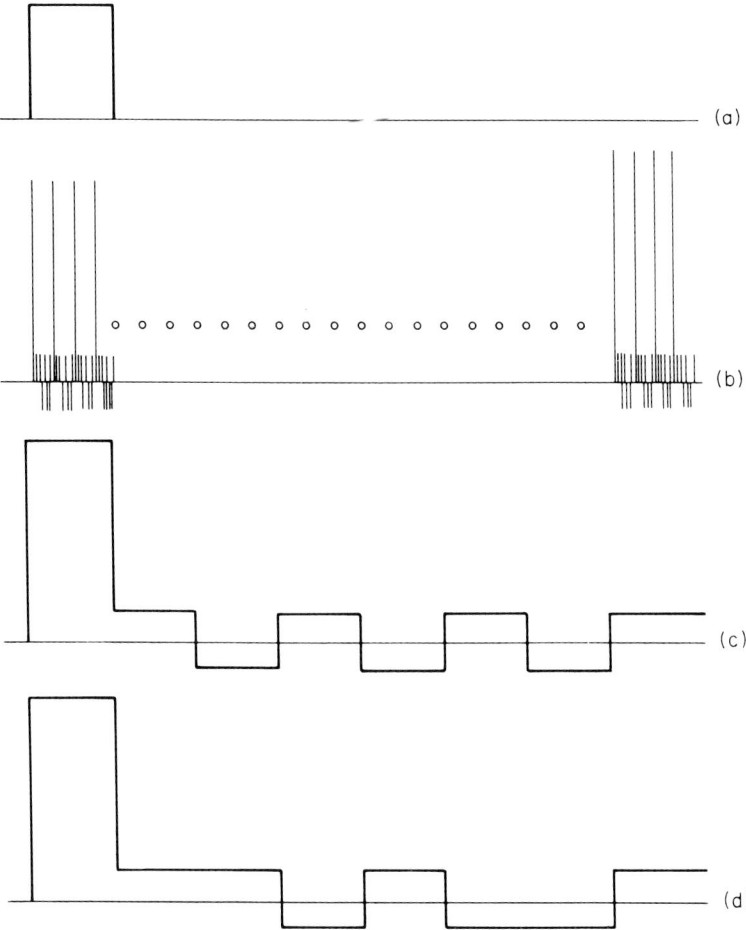

Fig. 4. Ordering of spectral coefficients. (a) Input function. (b) Unordered Walsh transform. (c) Sequency ordered Walsh transform. (d) Lexicographic ordered Walsh transform.

vector, and the resolution, $N = p^n$, of the input vector. Such information is sufficient to define the necessary parameters for implementation of a large class of transformations, all in $pN \log_p N$-operations.

It is instructive at this point to mention that the output of the analyzer consists of the coefficients of the orthogonal waveforms defining the vectors in the matrix transformation. However, the ordering of the coefficients is not necessarily in the most convenient format as witnessed by the reordering requirement in the fast Fourier transform literature. While no fixed reordering will be desirable for all classes of transformations, two specific permutations will be described due to their usefulness in observing Fourier spectra and Walsh spectra.

TABLE I

Case	Reference	p by p orthogonal submatrices	Orthogonal transform
1	Constant	$M_r \neq M_s \; \forall \, r, s$	Kronecker transform, Eq. (9)
2	Constant	$M_r = M_s \; \forall \, r, s$	Kronecker transform, Eq. (10)
3	Arbitrary	Case 1 or 2	Orthogonal transform
4	Constant	$\begin{bmatrix} \cos\theta & \sin\theta \\ \sin\theta & -\cos\theta \end{bmatrix}$	Eq. (18b) $\theta = 0°$, diagonal transform $\theta = 45°$, Walsh Hadamard transform Eq. (19a) $\theta = 90°$, opposing diagonal transform
5a	$g(x)^a$	$\left[\exp\dfrac{j2\pi ux}{p} \right]$	$c = 1$, Fourier transform with $N = p^n$ $c = 0$, generalized Walsh transform, Eq. (21b)
5b	$g(x)^a$	$\begin{bmatrix} 1 & 1 \\ 1 & -1 \end{bmatrix}$	$c = 1$, Fourier transform with $N = 2^n$ $c = 0$. Walsh Hadamard transform, Eq. (19a)

[a] $g(x) = \exp\left\{ j\dfrac{2\pi ckl}{p^n} \right\}$
where
$\quad x = kp^{n-1} + l, \quad c = $ real number
and
$\quad k = 0, 1, \ldots, p - 1, \quad l = 0, 1, \ldots, p^{n-1} - 1$

5.2 GENERALIZED SPECTRAL ANALYSIS   93

The first reordering is the one most commonly used in fast Fourier techniques of resolution equal to a power of 2 and consists in decifering the nth spectral coefficient in terms of its reverse binary representation. Actually, this technique is an application of the principle of decifering the nth generalized spectral coefficient in terms of its reverse lexicographic or dictionary representation as given earlier in Eq. (6a) and (6b). Such a method reorders any Kroneckered transformation and for a Fourier transform of resolution equal to a power of p, the spectral coefficients are reordered according to increasing frequency. The second reordering is useful for the Walsh Hadamard transforms and can best be described in terms of sequencies, a term introduced by Harmuth and discussed in the study of digital Walsh filters [23]. The sequency of a row of the Hadamard or Walsh matrix is the number of zero crossings within that row. As can be seen from Eq. (25), the H_3 matrix has rows with sequencies ranging from 0 to 7. However, they are unordered in terms of increasing sequency. Therefore, a useful ordering at the output of the spectrum analyzer for Walsh transforms places the coefficients in terms of increasing sequency. This ordering has been used in two-dimensional Hadamard image processing [16]. An example comparing the unordered, lexicographic ordering, and sequency ordering is presented in Fig. 4.

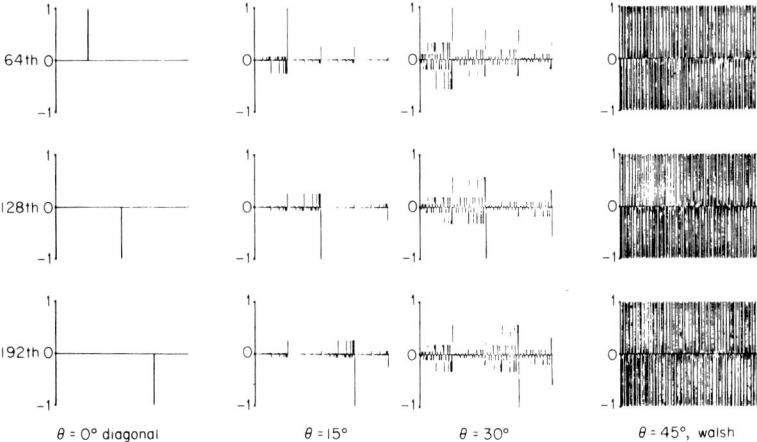

Fig. 5. Transition from diagonal to Walsh.

94 ORTHOGONAL TRANSFORMATIONS

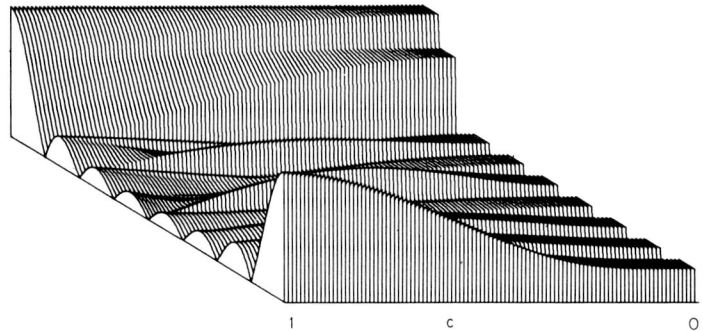

Fig. 6a. Orthogonal decomposition of a shifted block pulse (Fourier to Walsh transition).

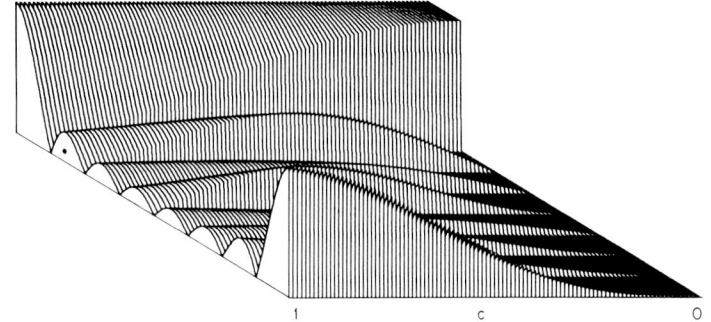

Fig 6b. Orthogonal decomposition of a block pulse (Fourier to Walsh transition).

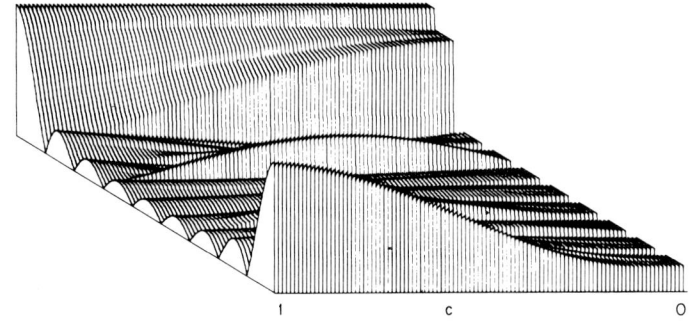

Fig. 6c. Orthogonal decomposition of a shifted block pulse (Fourier to generalized Walsh set of order 3).

5.2 GENERALIZED SPECTRAL ANALYSIS **95**

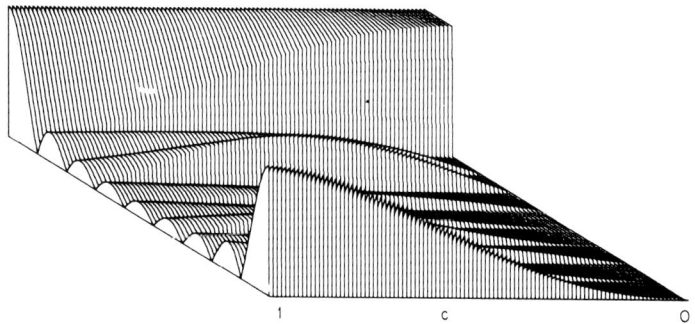

Fig. 6d. Orthogonal decomposition of a block pulse (Fourier to generalized Walsh Set of order 3).

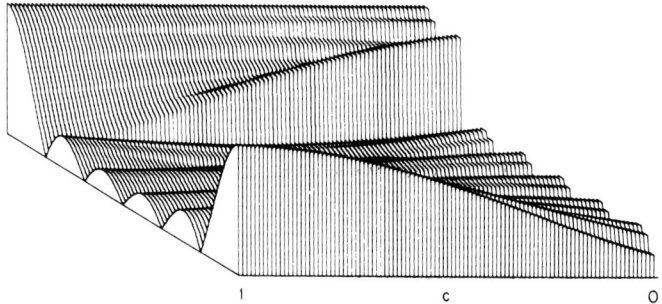

Fig. 6e. Orthogonal decomposition of a shifted block pulse (Fourier to generalized Walsh set of order 6).

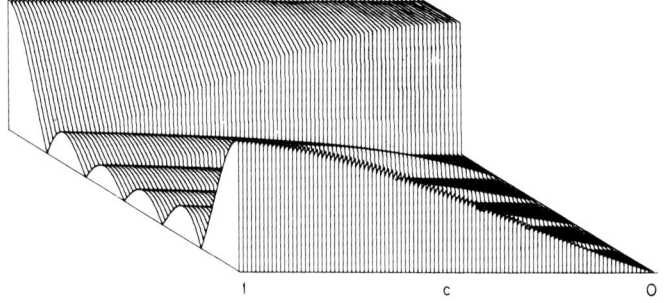

Fig. 6f. Orthogonal decomposition of a block pulse (Fourier to generalized Walsh set of order 6).

Returning now to some applications of the general spectrum analyzer, consider again the block diagram of Fig. 3. From a core matrix and reference waveform parameter viewpoint we can express a few examples of possible orthogonal transformations in Table I. Note that when $c = 0$ then $g(x)$ becomes the constant value 1. Examples from Case 4 and 5 will now be presented.

In Case 4, it is evident that when the core matrix ranges over θ, the energy on any one orthogonal vector starts totally concentrated at the diagonal point for $\theta = 0$ degrees and ranges to a uniform energy distribution for $\theta = 45$ degrees (Walsh–Hadamard case). Figure 5 presents some waveforms of the transform for various values of θ from 0 to 45 degrees. The resolution of the orthogonal matrix is $N = 2^8$. Decomposing input vectors into coefficient of such orthogonal functions has found application in two-dimensional image processing [17].

Case 5 from the table presents a class of orthogonal transformations which range from the Walsh or generalized Walsh transform to the Fourier transform as c increases from 0 to 1. Examples of such transformations are shown in Fig. 6 for three different resolutions 2^8, 3^5, 6^3 and block pulse waveform inputs. The lexicographic reordering technique is used in all transforms thus resulting in the split sin x/x pattern for the Fourier cases. Figures 6a and 6b are the transforms ranging from the Fourier to the Hadamard for a square wave in two different positions respectively. Figure 6c and 6d are the transformations ranging from the Fourier to the generalized Walsh of order 3, and Figs. 6e and 6f are for the Fourier to generalized Walsh of order 6 transition. In all cases the displays are the magnitude of the transforms and by the inner product preserving property of orthogonal transforms, the integrated area under each display is constant (graphical normalization may make this statement invalid).

It is instructive to review the transformations presented thus far for comparison purposes. A variety of orthogonal transformations have been presented which are implementable in relatively efficient fashion on a digital computer. However, the number of parameters defining the transformations allows for an extremely large number of orthogonal decompositions. Consequently, a comparison of some of the transformations for application purposes becomes desirable. One such tool for such an effort is afforded by an quantum analysis of the

5.2 GENERALIZED SPECTRAL ANALYSIS

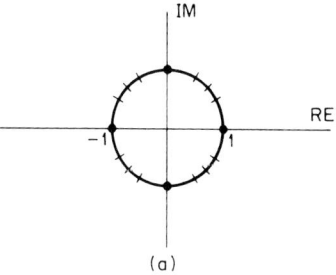

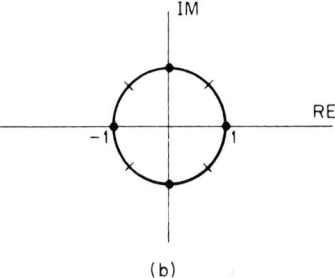

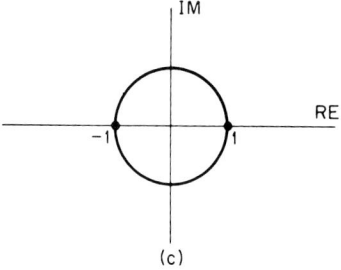

Fig. 7. Quantum levels on unit circle. (a) Fourier transform; N distinct quantum levels. (b) Generalized Walsh transform; p distinct quantum levels. (c) Walsh–Hadamard transform; ± 1, 2 distinct quantum levels.

transformations in question. This is done for the Fourier, generalized Walsh, and Hadamard-Walsh transforms in Fig. 7. It is evident from the figure that the Fourier transform has the largest number of quantum values $N = p^n$, while the Hadamard transform has the least, 2. It is also evident that the number of quantum levels in any orthogonal vector of the transformation matrix is greatest for the Fourier and least for the Hadamard. Algorithmically, the fewer the quantum levels, the more efficient the implementation of the transformation. Consequently, in the Fourier transformation operation a total of $N = p^n$-different levels must be generated and stored for eventual multiplication whereas in the Hadamard transform, only two levels, ± 1, need be generated and stored for multiplication. However, in this case, generation, storage, and multiplication reduce simply to addition and subtraction. In addition, the number of quantum levels in a transformation matrix is an indication of the effective quantization performed on a data input vector when correlating with the individual columns of the matrix in the orthogonal decomposition process. Consequently, if the input is only quantized to a given number of levels it is inefficient to correlate it with vectors of far greater quantization resolution especially if additional computational costs are incurred. These considerations are probably more significant in the design of special purpose digital analyzers but still could become factors in efficient general purpose implementation.

5.3 PATTERN RECOGNITION APPLICATIONS

While the applications of orthogonal transformations span a wide spectrum of interests, particular attention will be focused on the use of a generalized spectral analysis as a research tool in the emerging field of pattern recognition. The pattern recognition task is one which is conceptually quite easy to describe [24]. Sensory or unprocessed observations are obtained through some transducer system providing, usually, large quantities of raw data which comprise a pattern space, P. Quite often such data is indiscriminantly gathered, due to a lack of adaptive or recognition device, necessitating a method or technique of extracting pertinent features and combining correlated data to implement a bandwidth reduction, the resulting data of which will describe a feature space F. Finally the significant information in the feature space will then be used to form the classification space, C, in which the original pattern will have been properly classified. Consequently techniques are necessary to provide the transitions, probably noninvertible, from the pattern through the feature to the classification space $P \to F \to C$. If each sample of the pattern space is assumed to be a dimension whose range is the real axis or complex plane, then the $P \to F$ transition becomes a dimensionality reduction problem and digital computers are most often used for implementation purposes. This phase of the pattern recognition process provides the feature extraction and feature selection to define the F space. While a large number of pattern recognition techniques supplement this process with human experience, it is most desirable to develop an independent adaptive or learning scheme which does the feature extraction automatically, especially for recognition tasks which are not augmentable by human observation. Thus it is desirable to have the feature selection process be a function only of the learning algorithm, a suggestion of which follows.

Consider an N-dimensional pattern recognition problem where it is very probable that all N dimensions are not linearly independent but it is unknown which dimensions are more relevant and should be retained in a dimensionality reduction attempt. Let there be K possible classifications and the recognition device is to "learn" a classification procedure based on K prototype vectors P_k, each of dimension N and each belonging to a different one of the K classes. The learning procedure is to be developed by maximizing the variances of the prototype data over a variety of rotated vector spaces in the search for the fewest number of rotated dimensions with which proper classification can still be made. However, this technique is equivalent to performing an orthogonal transformation, a generalized spectral analysis, which maps the pattern space to a rotated pattern space in which the new basis vectors or dimensional axes are no longer block pulses of adjacent positions but are the vectors comprising the orthogonal matrix. Such a transformation represents a rotation of coordinates of an N-dimensional coordinate system into a new set of orthogonal coordinates also of N dimensions. Consequently, it seems reasonable that for a given pattern recognition problem, defined by its prototypes, there exists a coordinate rotation that would cause certain dimensions to become more relevant than others thereby allowing a classification decision to be based on fewer but more relevant dimensions. Such a technique is also known as principle component analysis [25-30].

The question of resolving the most relevant dimensions in a given rotational transformation space can be approached by maximizing the variances of the rotated dimensions. Boulton has attacked this problem specifically for the Walsh or Hadamard transform and he has suggested the concept of an "activity" criteria in which the activity of a given dimension in the transform domain is the variance of that coefficient taken over all K rotated prototypes [31]. Thus for any given transformation, N variance measures, σ_i^2, are obtained, one for each dimension in the transformed space. If the coordinates are ordered according to their variances, then the dimensions with the M largest variances becomes the M dimension most relevant. Note that usually $M \ll N$ and the dimensionality reduction is given by the ratio N/M. In matrix notation the vector prototypes P_κ are transformed into K rotated prototypes V_k and the M most active elements in the rotated prototypes are retained for classification purposes. Thus

$$P_k[T_l] = V_k$$
$$\sigma_i^2 = E\{(V_i - E\{V_i\})^2\}, \qquad i = 1, \ldots, N$$

where $E\{\ \}$ is the expectation operator taken over all K prototypes for each rotated dimension indexed by i. The set of M most active elements or dimensions is defined as the maximum variance set, S.

While the technique described above will provide a dimensionality reduction, it has not yet been defined as to which orthogonal transformation is the best rotation. If the allowable rotations are limited to those implementable in the $pN \log_p N$ algorithm presented earlier, then a suboptimum best transformation can be selected from this class. The maximum variance set $S(l)$ now becomes a function of the lth rotation and the transformation selection process is to minimized $S(l)$ over all l. Initially it would probably be desirably to restrict the orthogonal transformations to those with constant magnitude at each element in the rotation matrix. In this way no dimension is favored *a priori*. Thus the Fourier, Walsh, generalized Walsh, and all Hadamard transforms are immediate candidates for implementation.

The result of the method described above will be a feature space, F, composed of relatively few dimensions but each sample theoretically has a greater entropy than those in the pattern space P. Traditional classification procedures can now be invoked in the final recognition process based not on the N dimensions of the pattern space, but on the M dimensions of the feature space.

5.4 CONCLUSIONS

This chapter on orthogonal transformations is oriented specifically toward digital computer implementation. When in a computer environment, it is not necessary to restrict one's focus of attention to only the Fourier operations although they have a strong correspondence with the optical data processing field. Due to the general nature of digital processing, it is possible to decompose functions or images into a wide variety of orthogonal transformations that have little or no resemblence to physical phenomena in the real world. However, this does not preclude their possible usefulness. Thus, this chapter is devoted to the search for a class of rapidly and efficiently implementable orthogonal transformations for general purposes of image decomposition. While it is not expected that an efficient technique will be found for all eigenvector decompositions of real systems, it is suggested that for certain applications, the fast transformations described herein will find considerable use. The redundancy in the definition of Kronecker matrices is utilized to define an efficient relation for Kronecker products. These Kronecker products are then equated to a matrix factorization into matrices of high density zero-element content. This then allows an efficient implementation of a vector-matrix multiplication in fewer than the normally required operations. The fast algorithms derived are particularly appealing for they implement certain well-known orthogonal transformations: the Fourier, Walsh, Hadamard, and Haar transforms specifically. The concepts of generalized spectral analysis are laid down and numerous examples are presented displaying the large number of possibilities of transformations using the fast techniques derived earlier. A brief quantum value analysis is present for possible comparison purposes between transformations. Finally a brief section is included on the feature selection applications of orthogonal transformations in the pattern recognition environment. While no experimental work is presented on this particular topic, it is suggested that the large class of transforms presented in this chapter will allow for a means of implementing automatic feature selection procedures. With the tools presented in this chapter, it is now possible to study image coding in a digital communication environment by decomposing images into a variety of two-dimensional orthogonal transformations. The following chapters pursue this subject in some detail.

REFERENCES

1. J. W. Cooley and J. W. Tukey, "An Algorithm for the Machine Calculation of Complex Fourier Series," *Math. Computation* **19**, 297-301 (April 1965).
2. M. C. Pease, "An Adaptation of the Fast Fourier Transform for Parallel Processing," *J. Assoc. Computing Machinery* **15**, No. 2, 252-264 (April 1968).
3. *IEEE Trans. Audio Electroacoustics*, special issue on Fast Fourier Transforms, **Aa-15** (June 1967).
4. W. T. Cochran *et al.* "What is the Fast Fourier Transform," *Proc. IEEE* **55**, No. 10, 1664 (1967).
5. *IEEE Trans. Audio Electroacoustics*, special issue on Fast Fourier Transforms, **Aa-17** (June 1969).
6. H. C. Andrews, "A High Speed Algorithm for the Computer Generation of Fourier Transforms," *IEEE Trans. Computers* **C-17**, No. 4, 373 (1968).
7. A. Papoulis, *Systems and Transforms with Application in Optics*, p. 49. McGraw-Hill, New York (1968).
8. A. Papoulis, *Systems and Transforms with Applications in Optics*, p. 152. McGraw-Hill, New York, 1968.
9. D. Slepian and H. O. Pollak, "Prolate Spheroidal Wave Functions, Fourier Analysis and Uncertainty—I," *BSTJ* **40**, 43-63 (1961).
10. W. B. Davenport, Jr. and W. L. Root, *An Introduction to the Theory of Random Signals and Noise*. McGraw-Hill, New York, 1958.
11. H. C. Andrews and J. Kane, "Kronecker Matrices, Computer Implementation, and Generalized Spectra," *J. Assoc. Computing Machinery* (April 1970). Vol. 17, No. 2, pp. 260-268.
12. I. J. Good, "The Interaction Algorithm and Practical Fourier Analysis," *J. Roy. Statist. Soc. (London)* **B20**, 361 (1958).
13. J. E. Whelchel, Jr. and D. F. Guinn, "The Fast Fourier-Hadamard Transform and Its Use in Signal Representation and Classification," *Eascon 1968 Convention Record* pp. 561-573 (1968).

14. R. Bellman, *Introduction to Matrix Analysis.* McGraw-Hill, New York, 1960.
15. J. L. Walsh, "A Closed Set of Normal Orthogonal Functions," *Ann. J. Math.* **45,** 5-24 (1923).
16. W. K. Pratt, J. Kane, and H. C. Andrews, "Hadamard Transform Image Coding," *Proc. IEEE* **57,** No. 1, 58-68 (January 1969).
17. H. C. Andrews, and W. K. Pratt, "Transform Image Coding," Polytechnic Institute of Brooklyn *Internat. Symp. Computer Processing in Communications. April 1969.* Polytechnic Instit. Brooklyn, New York, 1969.
18. H. E. Chrestenson, "A Class of Generalized Walsh Functions," *Pacific J. Math.* **5,** 17-31 (1955).
19. A. Haar, "Zur Theorie der Orthogonalen Funktionen-Systeme," Inaugural Dissertation, *Math. Ann.* **69,** 331-371 (1910).
20. G. Alexits, *Convergence Problems of Orthogonal Series* **62**. Pergamon Press, New York, 1961.
21. C. Watari, "A Generalization of Haar Functions," *Tohoku Math. J.* **8,** 286-290 (1956).
22. H. C. Andrews and K. Caspari, "A Generalized Technique for Spectral Analysis," *IEEE Trans. Computers* **C-19,** No. 1, 16-25 (Jan., 1970).
23. H. F. Harmuth, "Sequency Filters Based on Walsh Functions," *IEEE Trans. Electromagnetic Compatibility* **EMC-10,** No. 2, 293-295 (1900).
24. H. J. Bremermann, "Pattern Recognition, Functionals, and Entropy," *IEEE Trans. Bio-Med. Eng.* **BME-15,** No. 3, 201-207 (July 1968).
25. S. S. Wilks, *Mathematical Statistics.* Wiley, New York, 1962.
26. C. R. Rao, "The Use and Interpretation of Principle Components Analysis in Applied Research," *Sankhyā, Indian I Stat. Ser. A* **26,** 329-359 (1964).
27. S. Watanabe, "Karhunen-Loeve Expansion and Factor Analysis, Theoretical Remarks and Applications," *Proc. Conf. Information Theory, 4th Prague 1965.*
28. J. Raviv and D. N. Streeter, "Linear Methods for Biological Data Processing," *IBM Res. Rep.* RD-1577 (1965).
29. J. A. McLaughlin and J. Raviv, "Nth Order Autocorreleations in Pattern Recognition," *Inform. Control* **12,** No. 2, 121-142 (February 1968).
30. V. R. Algazi and D. J. Sakrison, "On the Optimality of the Karhunen-Loeve Expansion," *GIT* **IT-15,** No. 2, 319 (March 1969).
31. P. I. P. Boulton, "Smearing Techniques in Pattern Recognition," Ph. D. Thesis, University of Toronto, 1966.

Chapter 6 IMAGE TRANSFORMS[1]

6.0 INTRODUCTION

The classic problem in the design of image coding systems for digital communication links is the search for a coding method which will minimize the number of code symbols required to describe an image. This coding method must not degrade the quality of the image beyond certain fidelity limits, and furthermore, the coding method must not be overly sensitive to channel errors. A great amount of investigation has been performed in the search for such image coding systems [1-3]. Unfortunately, most of the systems developed either do not exhibit satisfactory performance, or are too difficult to implement. The transform image coding method discussed in this chapter is a new approach to the problem of image coding. This image coding system achieves a reasonably large bandwidth reduction and offers a certain immunity to channel errors without significant image degradation as will be demonstrated in the following chapter.

The introduction of the fast Fourier transform algorithm [4-8] has led to the investigation of the Fourier transform image coding technique whereby the two-dimensional Fourier transform of an image is transmitted over a channel rather than the image itself [9-12]. This investigation has itself led to the study of a related image coding technique in which an image is transformed by a Hadamard matrix

---

[1] This chapter is by Harry Andrews and William Pratt.

operator [13-15]. The Hadamard matrix is a square array of plus and minus ones whose rows and columns are orthogonal to one another. A high speed Hadamard transform computational algorithm, similar to the fast Fourier transform algorithm has been developed [13].

The Fourier and Hadamard transforms are but two of a large number of transforms that have potential applications for image coding. In this chapter, the general properties of image transforms are considered. Specific experimental examples of the performance of the Fourier and Hadamard image transforms are given.

Figure 1 illustrates the block diagram of a generalized transform image coding system. In this system a transform is performed on the intensity samples of the image. The image transform samples are then quantized and coded for transmission over a digital link. At the receiver the received data is decoded, and an inverse transform is performed to reconstruct the original image. In principle the transforms could be implemented by optical, electrical, or digital techniques. The experimental results presented in this chapter have been obtained for a general purpose digital computer implementation of the image transforms. No attempt has been made to determine the "best" means of transform implementation, other than to present the most efficient computer algorithms.

As a prelude to the subsequent sections, Figs. 2-7 contain examples of the two-dimensional Fourier and Handamard transforms of an image. The original images, containing 256 by 256 elements and linearly quantized to 64 grey levels, have been transformed on a general purpose computer. Spatial and transform domains have been displayed on a cathode ray tube monitor for photographic recording. All transforms contain 256 by 256 sample points.

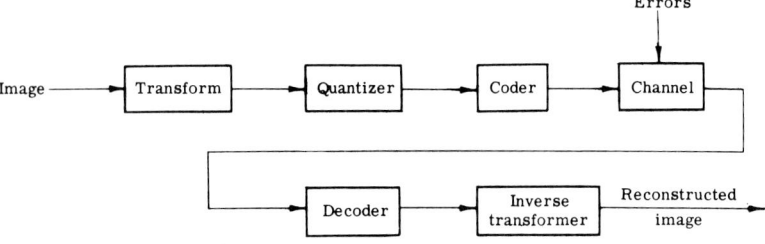

Fig. 1. Image transform coding system.

6.0 INTRODUCTION 107

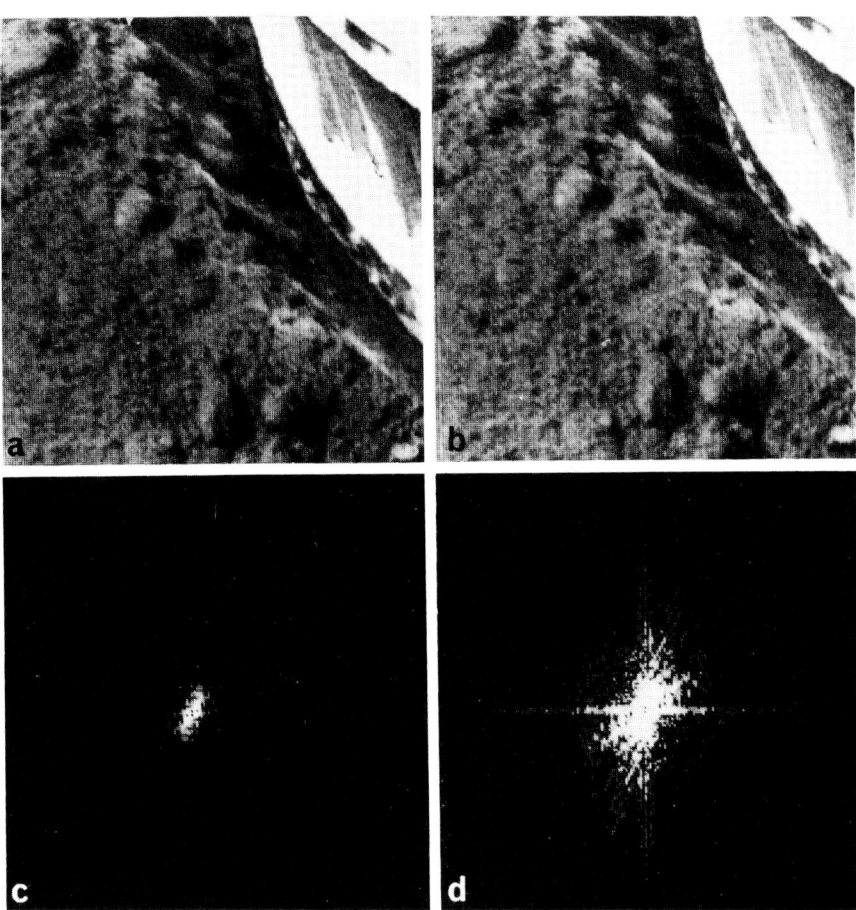

Fig. 2. Fourier transforms of surveyor spacecraft footpad. (a) Original. (b) Inverse Fourier transform of Fourier transform. (c) Magnitude of Fourier transform. (d) Logarithm of magnitude of Fourier transform.

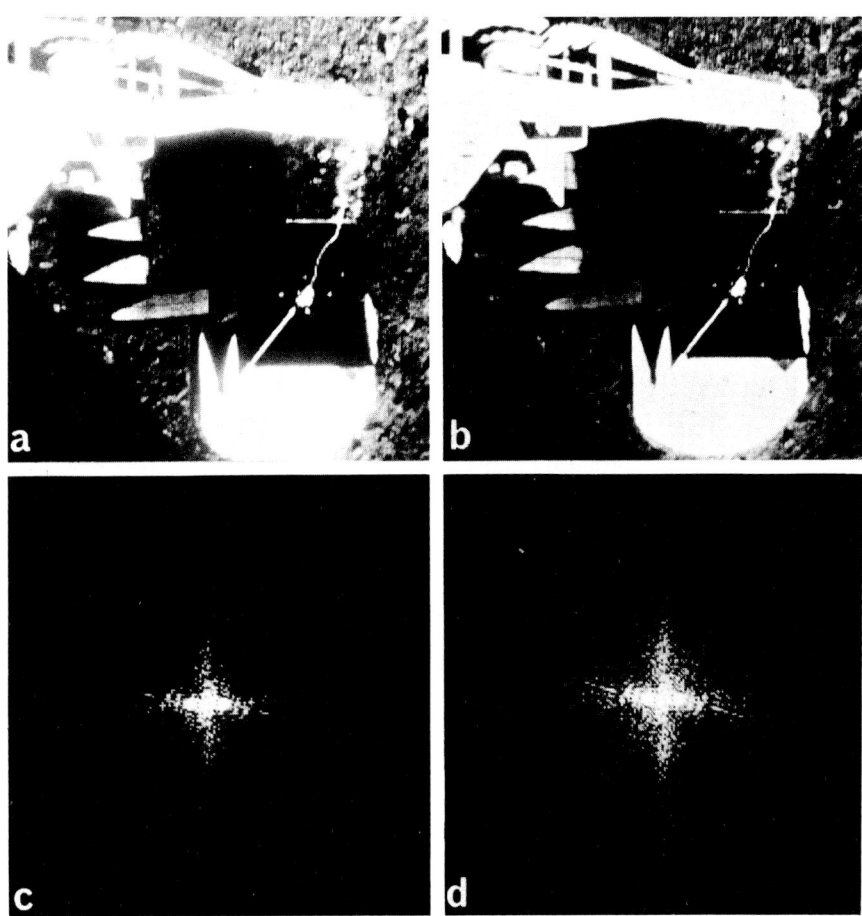

Fig. 3. Fourier transforms of surveyor spacecraft experimental box. (a) Original. (b) Inverse Fourier transform of Fourier transform. (c) Magnitude of Fourier transform. (d) Logarithm of magnitude of Fourier transform.

6.0 INTRODUCTION 109

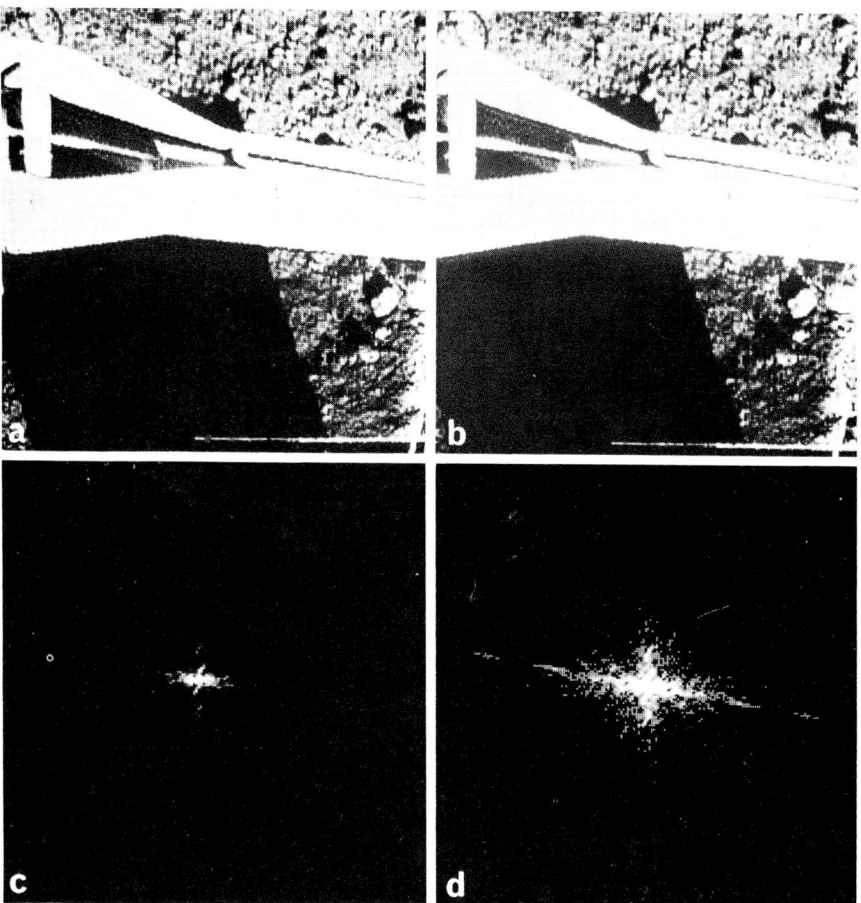

Fig. 4. Fourier transforms of surveyor spacecraft boom. (a) Original. (b) Inverse Fourier transform of Fourier transform. (c) Magnitude of Fourier transform. (d) Logarithm of magnitude of Fourier transform.

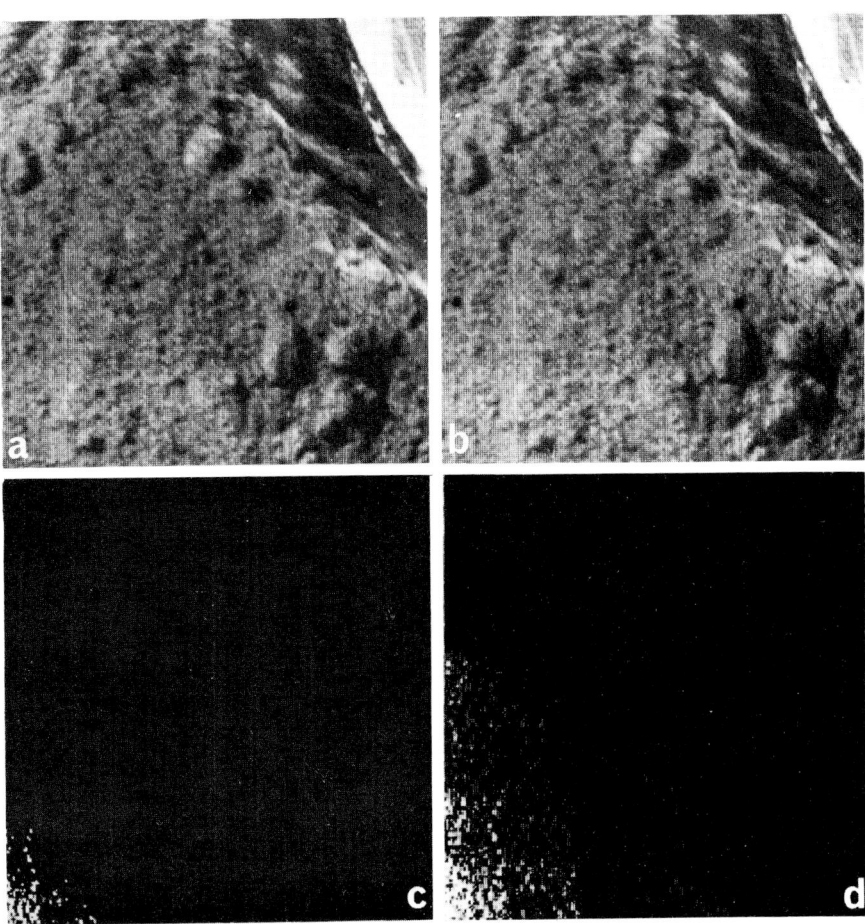

Fig. 5. Hadamard transforms of surveyor spacecraft footpad. (a) Original. (b) Hadamard transform of Hadamard transform. (c) Magnitude of Hadamard transform. (d) Logarithm of magnitude of Hadamard transform.

6.0 INTRODUCTION 111

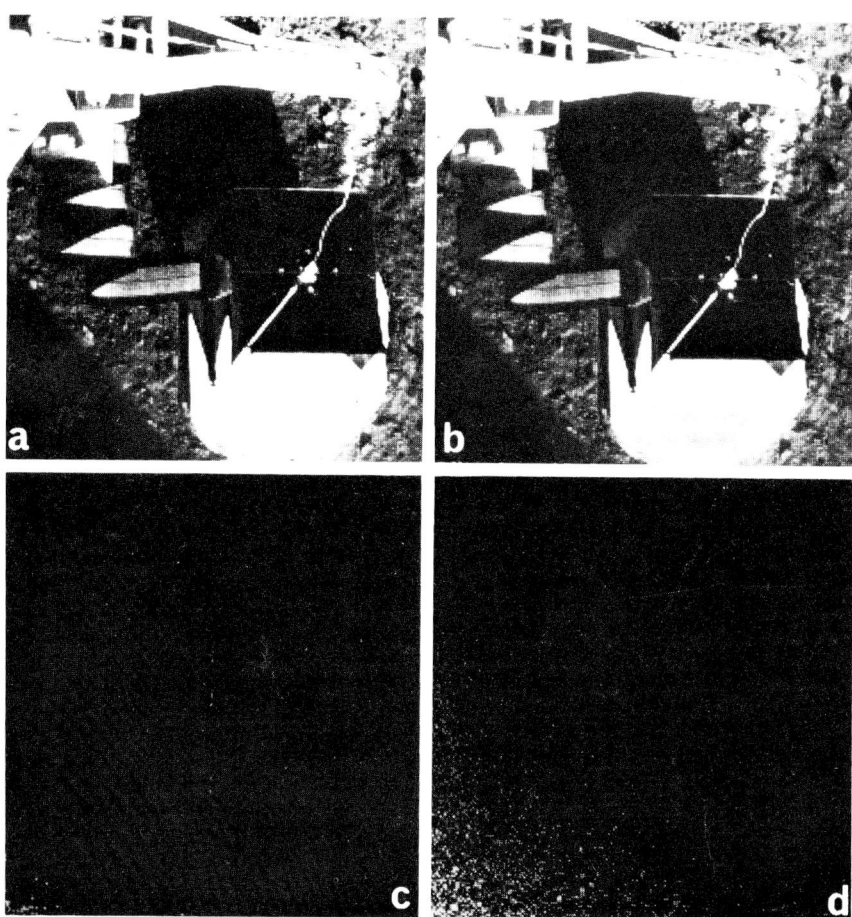

Fig. 6. Hadamard transforms of surveyor spacecraft experimental box. (a) Original. (b) Hadamard transform of Hadamard transform. (c) Magnitude of Hadamard transform. (d) Logarithm of Magnitude of Hahamard transform.

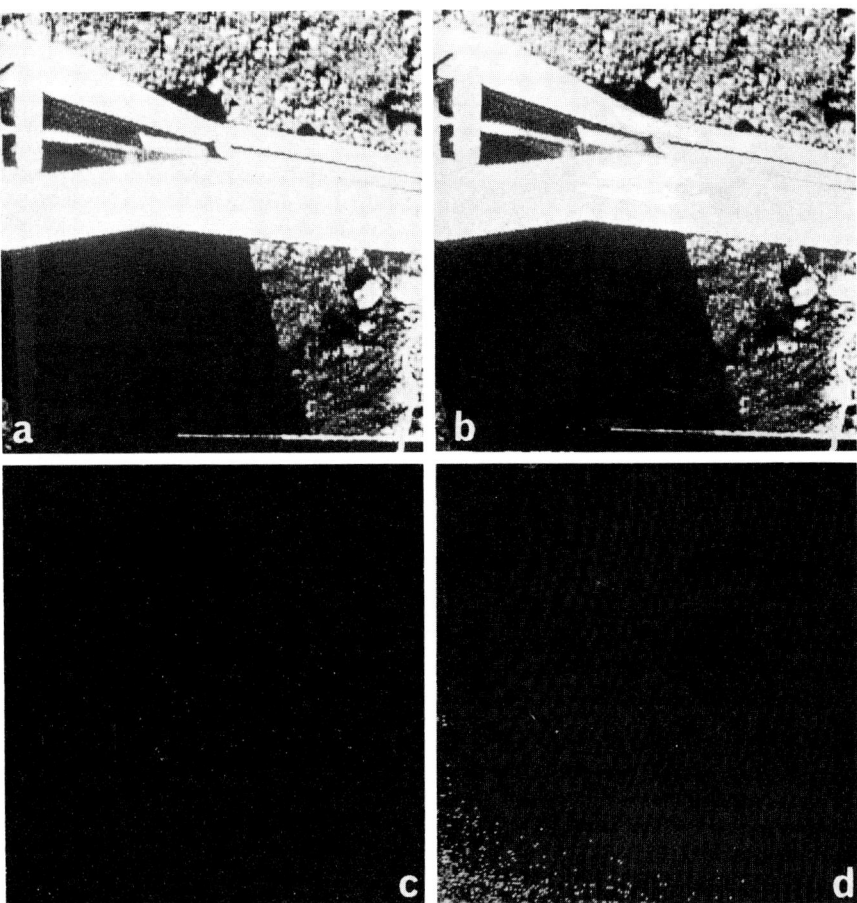

Fig. 7. Hadamard transforms of surveyor spacecraft boom. (a) Original. (b) Hadamard transform of Hadamard transform. (c) Magnitude of Hadamard transform. (d) Logarithm of magnitude of Hadamard transform.

Figures 2-7 indicate that there is no apparent image degradation between the originals and the double transforms as a result of the image transform operations. The inherent correlation of elements in the originals has caused the energy in the transform domains to be squeezed toward the zero spatial frequencies. It is this characteristic of transformed images that is exploited to achieve a bandwidth reduction. The error immunity property of transform coding results from the inherent averaging operation of the transform. Each intensity sample of a reconstructed image is a weighted function of all transform samples. Hence, the magnitude of a single channel error is distributed over all of the reconstructed elements. In addition, as Chapter 7 demonstrates, it will be possible to error correct code a small percentage of transform samples to achieve even greater noise immunity.

6.1 IMAGE TRANSFORMATION

Consideration is given in this section to the mathematical formulation of image transforms. The characteristics and properties of the Fourier, Hadamard, and other transforms are then developed, and finally conditions are given for the existence of fast computational algorithms.

6.1.1 Formulation

An original image may be represented by an array of intensity components or samples over the image surface by two dimensional sampling. In the transform coding system, it is conceptually possible to process the entire image or subsections of the image. The "best" image or subsection size is dependent upon the degree of spatial correlation of the image and the amount of processing permitted. For the present discussion an image array will be considered to be a square array of N^2 intensity samples described by the function, $f(x, y)$ over the image coordinates (x, y). Then the two-dimensional forward transform of the image array, $F(u, v)$, itself defined on a square array of N^2 points, may be expressed as

$$F(u, v) = \sum_{x=0}^{N-1} \sum_{y=0}^{N-1} f(x, y)\, a(x, y, u, v) \qquad (1)$$

where $a(x, y, u, v)$ is the forward transformation kernel. The kernel is said to be separable if it can be written as

$$a(x, y, u, v) = a_1(x, u) a_2(y, v) \tag{2}$$

A separable two-dimensional transform can be computed in two steps. First, a one-dimensional transform is taken along each row of the image, $f(x, y)$, yielding

$$F(u, y) = \sum_{x=0}^{N-1} f(x, y) a_1(x, u) \tag{3}$$

Next, a second one-dimensional transform is taken along each column of $F(u, y)$ giving

$$F(u, v) = \sum_{y=0}^{N-1} F(u, y) a_2(y, v) \tag{4}$$

The transformation kernel is called separable symmetric if

$$a(x, y, u, v) = a_1(x, u) a_1(y, v) \tag{5}$$

For ease of implementation, the separable symmetric property is desirable. Furthermore, since the statistical intensity variations of most images are nearly the same in the vertical and horizontal directions only separable symmetric kernels usually need be considered.

A reverse transform may be defined as

$$f(x, y) = \sum_{u=0}^{N-1} \sum_{v=0}^{N-1} F(u, v) b(x, y, u, v) \tag{6}$$

where $b(x, y, u, v)$ is the reverse transformation kernel.

It is often useful to express two-dimensional transforms in matrix notation. For a transform kernel that is separable symmetric let:

$$[f] = \text{image matrix} \tag{7}$$
$$[F] = \text{transformed image matrix} \tag{8}$$
$$[A] = \text{transform matrix} \tag{9}$$

Then by matrix multiplication

$$[F] = [A][f][A] \tag{10}$$

Now pre- and post-multiplication of each side of $[F]$ by a reverse transform matrix, $[B]$, gives

$$[\hat{f}] \equiv [B][F][B] = [B][A][f][A][B] \tag{11}$$

where $[\hat{f}]$ is in general an approximation of $[f]$. If the reverse transform matrix is the inverse matrix $[A]^{-1}$ of $[A]$, then

$$[\hat{f}] = [A]^{-1}[A][f][A][A]^{-1} \tag{12}$$

But

$$[A]^{-1}[A] = [A][A]^{-1} = [I] \tag{13}$$

where $[I]$ is the identity matrix. Hence

$$[\hat{f}] = [f] = [A]^{-1}[F][A]^{-1} \tag{14}$$

Thus, $f(x, y)$ and $F(u, v)$ can be expressed as two-dimensional transform pairs if $[A]$ has an inverse.

If $[A]$ is a unitary matrix, then by definition

$$[A]^{-1} = [A^*]^T \tag{15}$$

where $[A^*]$ is the complex conjugate matrix of $[A]$ and where $[A]^T$ is the matrix transpose of $[A]$. A real, unitary matrix is called an orthogonal matrix. For such a matrix

$$[A]^{-1} = [A]^T \tag{16}$$

Finally, if $[A]$ is a symmetric orthogonal matrix then

$$[A]^{-1} = [A] \tag{17}$$

If the forward transformation matrix, $[A]$, is constrained to be orthogonal, then the transformation can be interpreted as a decomposition of the image data into a generalized two-dimensional spectrum. Each spectral component in the transform domain corresponds to the amount of energy of that spectral orthogonal function within the original image. In this context the concept of frequency may now be generalized to include transformations of orthogonal functions other than sine and cosine waveforms. This type of generalized spectral analysis is useful in the investigation of specific orthogonal decompositions which are best suited for particular classes of images.

The following paragraphs contain an analysis of the Fourier, Hadamard, Kronecker, and Eigenvector transformations with particular emphasis on their applicability to image coding.

6.1.2 Fourier Transform

The discrete Fourier transform, with and without efficient computational algorithms, has long been used for signal analysis [6]. Only recently have Fourier transform methods been utilized for image coding [9-12]. The two-dimensional Fourier transform of an image field, $f(x, y)$, may be expressed as

$$F(u, v) = \frac{1}{N} \sum_{x=0}^{N-1} \sum_{y=0}^{N-1} f(x, y) \exp\left\{-\frac{2\pi i}{N}(xu + yv)\right\} \quad (18)$$

The inverse Fourier transform which reconstructs the original image is given by

$$f(x, y) = \frac{1}{N} \sum_{u=0}^{N-1} \sum_{y=0}^{N-1} F(u, v) \exp\left\{\frac{2\pi i}{N}(ux + vy)\right\} \quad (19)$$

Since the transform kernels are separable and symmetric, the two-dimensional transform can be computed as two sequential one-dimensional transforms.

The terms u and v are called the spatial frequencies of the image in analogy with time series analysis. When the Fourier transform relationship is expressed in the form given by Eq. (18), the origin, or zero spatial frequency term appears in the corner of the transform plane. For display purposes it is convenient to shift the origin to the center of the transform domain. This is easily accomplished by multiplying the image by the function $(-1)^{x+y}$ before the transformation. Let

$$G(u, v) = \frac{1}{N} \sum_{x=0}^{N-1} \sum_{x=0}^{N-1} (-1)^{x+y} f(x, y) \exp\left\{-\frac{2\pi i}{N}(ux + vy)\right\} \quad (20)$$

But since

$$(-1)^{x+y} = e^{i\pi(x+y)} \quad (21)$$

the function $G(u, v)$ may be written as

$$G(u, v) = \frac{1}{N} \sum_{x=0}^{N-1} \sum_{y=0}^{N-1} f(x, y)$$
$$\times \exp\left\{-\frac{2\pi i}{N}\left[\left(u - \frac{N}{2}\right)x + \left(v - \frac{N}{2}\right)y\right]\right\} \quad (22)$$

or

$$G(u, v) = F\left(u - \frac{N}{2},\ v - \frac{N}{2}\right) \quad (23)$$

Thus, the origin moves to the center of the transform domain.

Even though $f(x, y)$ is a real positive function, its transform, $F(u, v)$ is in general complex. Thus, while the image contain N^2 components, the transform contains $2N^2$ components, the real and imaginary or magnitude and phase components of each spatial frequency. However, since $f(x, y)$ is a real positive function, $F(u, v)$, exhibits a property of conjugate symmetry. To illustrate this property let

$$F(u, v) = \frac{1}{N} \sum_{x=0}^{N-1} \sum_{y=0}^{N-1} f(x, y)$$
$$\times \left\{\cos\left[\frac{2\pi}{N}(ux + vy)\right] - i \sin\left[\frac{2\pi}{N}(ux + vy)\right]\right\} \quad (24)$$

The Fourier transform can be divided into real and imaginary components as

$$F(u, v) = F_R(u, v) - iF_I(u, v) \quad (25)$$

where, since $f(x, y)$ is real,

$$F_R(u, v) = \frac{1}{N} \sum_{x=0}^{N-1} \sum_{y=0}^{N-1} f(x, y) \cos\left[\frac{2\pi}{N}(ux + vy)\right] \quad (26)$$

and

$$F_I(u, v) = \frac{1}{N} \sum_{x=0}^{N-1} \sum_{y=0}^{N-1} f(x, y) \sin\left[\frac{2\pi}{N}(ux + vy)\right] \quad (27)$$

The cosine is even in u and v, and the sine is odd in u and v, hence

$$F_R(u, v) = F_R(-u, -v) \quad (28)$$

and
$$F_I(u, v) = -F_I(-u, -v) \tag{29}$$
Consequently
$$F(u, v) = F^*(-u, -v) \tag{30}$$

Figure 8 illustrates the conjugate symmetry property of the Fourier transform when the zero spatial frequency term is located at the center of the transform plane. Samples in quadrants (1) and (3) are complex conjugates of one another as are samples in quadrants (2) and (4). This property is further illustrated by magnitude displays of the Fourier transforms shown in Figs. 2-4. As a result of the conjugate symmetry property of the Fourier transform, it is only necessary to transmit the samples of one half of the transform plane; the other half can be reconstructed from the half-plane sample transmitted.[2] Hence, the Fourier transform of an image can be described by N^2 data components.

The two-dimensional Fourier transform of an image is essentially a Fourier series representation of a two-dimensional field. For the

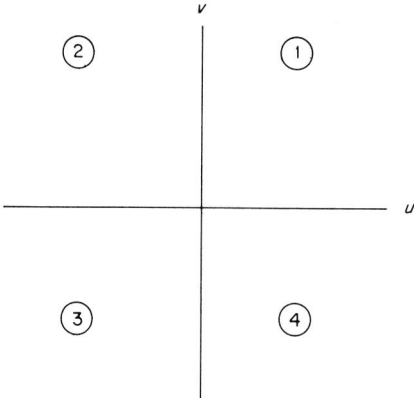

Fig. 8. Fourier domain quadrants.

---

[2] A reconstruction of the original can be obtained from the half-plane transform samples directly by a Hilbert filtering technique [9].

Fourier series representation to be valid the field must be periodic. Thus, the original image must be considered to be periodic horizontally and vertically as shown Fig. 9. The right-hand side of the image therefore abuts the left-hand side and the top and bottom of the image are adjacent. Spatial frequencies along the coordinate axes of the transform plane arise from these transitions. Although these are false spatial frequencies from the standpoint of being necessary for representing the image within the image boundary, they do not impair reconstruction. On the contrary, these spatial frequencies are required to reconstruct the sharp boundaries of the image.

The Fourier transform can be easily expressed in a matrix formulation by letting

$$\mathscr{W} \equiv \exp\left\{-\frac{2\pi i}{N}\right\}. \tag{31}$$

Then

$$[F] = [A][f][A]. \tag{32}$$

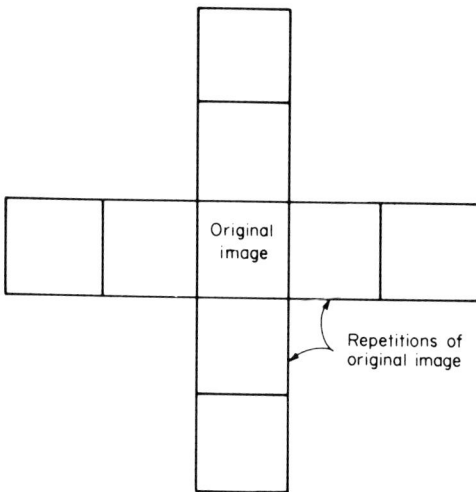

Fig. 9. Fourier series representation of an image.

where

$$[A] = \frac{1}{\sqrt{N}} \begin{array}{c} 0 \\ 1 \\ 2 \\ 3 \\ \vdots \\ \vdots \\ N-1 \end{array} \begin{array}{cccccc} 0 & 1 & 2 & 3 & \cdots & N-1 \\ \left[\mathcal{W}^0 \right. & \mathcal{W}^0 & \mathcal{W}^0 & \mathcal{W}^0 & \cdots & \mathcal{W}^0 \\ \mathcal{W}^0 & \mathcal{W}^1 & \mathcal{W}^2 & \mathcal{W}^3 & \cdots & \mathcal{W}^{N-1} \\ \mathcal{W}^0 & \mathcal{W}^2 & \mathcal{W}^4 & \mathcal{W}^6 & \cdots & \mathcal{W}^{2(N-1)} \\ \mathcal{W}^0 & \mathcal{W}^3 & & & & \\ \vdots & \vdots & & & & \\ \mathcal{W}^0 & \mathcal{W}^{N-1} & \cdots & \cdots & & \left. \mathcal{W}^{(N-1)^2} \right] \end{array} \begin{array}{c} x \\ \downarrow \end{array} \quad (33)$$

$u \longrightarrow$

The matrix is obviously symmetric. Computational simplification can be realized since

$$\mathcal{W}^{ux} = \mathcal{W}^{ux \bmod N} \qquad (34)$$

6.1.3 Hadamard Transform

The Hadamard transform is based upon the Hadamard matrix which is a square array of ± 1's whose rows and columns are orthogonal to one another [16–18]. If $[H]$ is an N by N Hadamard matrix, then the product of N and its transpose is

$$[H][H]^T = N[I] \qquad (35)$$

If H is a symmetric Hadamard matrix, then Eq. (35) reduces to

$$[H][H] = N[I] \qquad (36)$$

A Hadamard matrix multiplied by the normalization factor $1/N^{1/2}$ is an orthonormal matrix.

The lowest order Hadamard matrix is

$$[H_2] = \begin{bmatrix} 1 & 1 \\ 1 & -1 \end{bmatrix} \qquad (37)$$

It is known that if a Hadamard matrix of order N exists $(N > 2)$, then $N \equiv 0 \pmod 4$. The existence of a Hadamard matrix for every value of N satisfying this requirement has not been shown, but constructions are available for nearly all permissible values of N up to 200. The simplest construction is for a Hadamard matrix of order $N = 2^n$ where n is an integer. In this case if $[H_N]$ is a Hadamard matrix of order N, the matrix

$$[H_{2N}] = \begin{bmatrix} H_N & H_N \\ H_N & -H_N \end{bmatrix} \quad (38)$$

is a Hadamard matrix of order $2N$. Figure 10 contains several

$N = 2$ $\quad$ Matrix $\quad$ Sequency

$\begin{bmatrix} + & + \\ + & - \end{bmatrix}$ $\quad$ 0
$\quad$ 1

$N = 4$ $\quad$ Matrix $\quad$ Sequency

$\begin{bmatrix} + & + & + & + \\ + & - & + & - \\ + & + & - & - \\ + & - & - & + \end{bmatrix}$ $\quad$ 0
$\quad$ 3
$\quad$ 1
$\quad$ 2

$N = 8$ $\quad$ Matrix $\quad$ Sequency

$\begin{bmatrix} + & + & + & + & + & + & + & + \\ + & - & + & - & + & - & + & - \\ + & + & - & - & + & + & - & - \\ + & - & - & + & + & - & - & + \\ + & + & + & + & - & - & - & - \\ + & - & + & - & - & + & - & + \\ + & + & - & - & - & - & + & + \\ + & - & - & + & - & + & + & - \end{bmatrix}$ $\quad$ 0
$\quad$ 7
$\quad$ 3
$\quad$ 4
$\quad$ 1
$\quad$ 6
$\quad$ 2
$\quad$ 5

Fig. 10. Hadamard matrices of order $N = 2^n$.

Hadamard matrices of order $N = 2^n$. Another simple construction is possible if $[G_M]$ and $[H_N]$ are Hadamard matrices of orders M and N, respectively. Then there exists a Hadamard matrix of order $M \cdot N$ given by

$$[H_{M \cdot N}] = \begin{bmatrix} g_{11}H_N & g_{12}H_N & \cdots & g_{1M}H_N \\ g_{21}H_N & & & \vdots \\ g_{M1}H_N & \cdots & \cdots & g_{MM}H_N \end{bmatrix} \quad (39)$$

Other constructions are given by Paley [19], Williamson [20], and Baumert et al. [21]. The set of known Hadamard matrices is sufficiently numerous to satisfy almost all size requirements for image coding.

A frequency interpretation can be given to the Hadamard matrix generated from the core matrix of Eq. (37). Along each row of the Hadamard matrix the frequency is called the number of changes in sign. Harmuth has coined the word "sequency" to designate the number of sign changes [22]. Figure 10 gives the sequency interpretation for several Hadamard matrices of binary order. It is possible to construct a Hadamard matrix of order $N = 2^n$ that has frequency components at every integer from 0 to $N - 1$.

This frequency interpretation of the rows of a Hadamard matrix leads one to consider the rows to be equivalent to rectangular waves ranging between ± 1 with a subperiod of $1/N$ units. Such functions are called Walsh functions [23-27] and are further related to the Rademacher functions [28]. Thus, in this context the Hadamard matrix merely performs the decomposition of a function by a set of rectangular waveforms rather than the sine-cosine waveforms associated with the Fourier transform.

For symmetric Hadamard matrices of order $N = 2^n$, the two-dimensional Hadamard transform may be written in the series form

$$F(u, v) = \frac{1}{N} \sum_{x=0}^{N-1} \sum_{y=0}^{N-1} f(x, y)(-1)^{p(x,y,u,v)} \quad (40)$$

where

$$p(x, y, u, v) \equiv \sum_{i=0}^{N-1} (u_i x_i + v_i y_i)$$

The terms u_i, v_i, x_i, and y_i are the binary representations of u, v, x, and y respectively. For example

$$(u)_{\text{DECIMAL}} = (u_{n-1}u_{n-2}\cdots u_1u_0)_{\text{BINARY}} \tag{41}$$

where $u_i \in \{0, 1\}$. In Eq. (40), the summation in the exponent is performed modulo two. This representation of the Hadamard transform is for the Hadamard matrix in "natural" form as given by Eq. (38). Another series representation exists for a Hadamard matrix in "ordered" form in which the sequency of each row is larger than the preceding row. By this representation

$$F(u, v) = \frac{1}{N} \sum_{x=0}^{N-1} \sum_{y=0}^{N-1} f(x, y)(-1)^{q(x,y,u,v)} \tag{42}$$

where

$$q(x, y, u, v) \equiv \sum_{i=0}^{n-1} [g_i(u)x_i + g_i(v)y_i] \tag{43}$$

and

$$\begin{aligned}
g_0(u) &\equiv u_{n-1} \\
g_1(u) &\equiv u_{n-1} + u_{n-2} \\
g_2(u) &\equiv u_{n-2} + u_{n-3} \\
&\vdots \\
g_{n-1}(u) &\equiv u_1 + u_0
\end{aligned} \tag{44}$$

The two-dimensional Hadamard transform may be computed in either natural or ordered form with an algorithm analogous to the fast Fourier transform computer algorithm.

6.1.4 Karhunen–Loeve Transform

In the transform threshold sampling technique of bandwidth reduction, presented in Chapter 7, only those transform samples whose magnitudes are greater than a threshold level are coded. The optimum transform for minimizing the number of transform samples lying within a region while satisfying a mean square error criterion between the original and the reconstructed image is the Karhunen–Loeve transform [29–36]. This transform is composed of eigenvectors of the correlation matrix of the original image, or class of images, to be coded.

The Karhunen–Loeve transform is not in general separable. Hence, the original image must be regarded as a vector rather than a matrix. Let

$$[f(z)] \equiv [f(x, y_1), f(x, y_2), \ldots, f(x, y_N)] \quad (45)$$

be a row vector composed of lines of the original image examined in the normal "raster" pattern. The correlation matrix of the image is an N^2 by N^2 matrix of the form[3]

$$[R] = E\{z_i z_j\}, \quad i = 1, 2, \ldots, N; \quad j = 1, 2, \ldots, N \quad (46)$$

If the correlation matrix is not known, it can be estimated from an ensemble of original images. Let $[f_k(z)]$ represent the kth of n images from which $[R]$ is to be estimated. Then

$$[R] \approx \frac{1}{n} \sum_{k=1}^{n} [f_k(z)]^T [f_k(z)] \quad (47)$$

The forward Karhunen–Loeve transform is the orthogonal matrix composed of the eigenvectors of the correlation matrix arranged such that

$$[A]^T R [A] = \begin{bmatrix} \lambda_1 & & & 0 \\ & \lambda_2 & & \\ & & \ddots & \\ 0 & & & \lambda_{N^2} \end{bmatrix} \quad (48)$$

where $\lambda_1 \geq \lambda_2 \geq \cdots \geq \lambda_{N^2}$ are the eigenvalues of $[R]$ arranged in descending order. The Karhunen–Loeve transform, $F(w)$, of the original image is then

$$[F(w)] = [f(z)][A] \quad (49)$$

And the reverse transform is

$$[B] = [A]^T \quad (50)$$

If only the first M of the N^2 columns of $[A]$ are employed in the transform, i.e.,

---

[3] $E\{\cdot\}$ = expected or mean value of the function within the brackets.

$$[F_M(w)] = [f(z)][A_M] \quad (51)$$
$\underbrace{}_{\substack{1\times M \\ \text{matrix}}} \quad \underbrace{}_{\substack{1\times N^2 \\ \text{matrix}}} \underbrace{}_{\substack{N^2\times M \\ \text{matrix}}}$

then the mean square error, ε, is

$$\varepsilon = \sum_{k=M+1}^{N^2} \lambda_k \quad (52)$$

Since the λ_k are monotonically decreasing in value the error will be minimum for any M.

There are two major problems associated with the use of the Karhunen-Loeve transform for image coding. The first is that a great amount of computation must be performed. The correlation matrix must be estimated if it is not known. Next the correlation matrix must be diagonalized to determine its eigenvalues and eigenvectors. Finally, the transform itself must be taken. In general, there is no fast computational algorithm for the transform. The second difficulty with the Karhunen–Loeve transform is that the mean square error is not a valid error criterion for many types of images.

However, for those classes of images for which the mean square error criterion is valid, the Karhunen–Loeve transformation may find application as a standard for bandwidth reduction capability. Furthermore, if the image is broken up into smaller subsections the computation of the transform may prove feasible.

6.1.5 Computational Algorithms

A characteristic of great importance for an image transform is the existence of a fast computational algorithm of the type available for the Fourier and Hadamard transforms. For a fast algorithm to exist it is necessary that the transform be factorable into matrices containing many zero elements [37].

It is not possible to identify matrices that are factorable into matrices containing many zero elements. No algorithms exist for determining the best factorization of factorable matrices. Therefore, the only recourse in finding an efficient computation algorithm for an arbitrary transform matrix is to generate trial factorizations and compare them. This is why the synthesis procedure of Chapter 5 was developed.

6.2 ANALYSIS OF IMAGE TRANSFORMS

The development of efficient quantization and coding methods for image transforms requires an understanding of the statistic distribution of energy in the transform domain and bounds on the distribution of the energy. In this section a stochastic model of transform samples is developed, energy bounds are derived, and computation requirements to preserve the energy distribution are presented.

6.2.1 Statistical Analysis

A complete statistical description of the effects of a general transformation operator on an original image is not possible. However, considerable insight into the general statistical description can be obtained from the Fourier transform by the relation between the Fourier transform spatial frequency and the concept of the generalized frequency of a transform.

In the statistical analysis of the two-dimensional Fourier transform let $\phi(x', y')$ be a continuous two-dimensional wide sense stationary random process with a bounded and continuous power spectral density $D(u, v)$, where u and v are real. It is desired to observe the process over the two-dimensional window, $(-I, -I; I, I)$, and to sample the process at N^2 uniformly spaced points within the window of observation. A new process, $F_{N,I}(u, v)$ depending on both the window of observation and the sampling period within the window, is formed as follows:

$$F_{N,I}(u, v) = \frac{1}{N} \sum_{x,y=0}^{N-1} \sum \phi\left(\frac{xI}{N}, \frac{yI}{N}\right) \exp\left\{\frac{2\pi iI}{N}(ux + vy)\right\} \quad (53)$$

The variance of $F_{N,I}(u, v)$ may be expressed in terms of a covariance function, ρ, on the process, ϕ. Thus

$$\sigma^2_{F_{N,I}}(u, v) = \frac{1}{N^2} \sum_{\tau,T=0}^{N-1} \varepsilon_\tau \varepsilon_T (n - \tau)(N - T)$$

$$\times \rho\left(\frac{\tau I}{N}, \frac{TI}{N}\right) \cos\left[\frac{2\pi I}{N}(u\tau + vT)\right] \quad (54)$$

where τ and T are integer values representing the two dimensional shift in the sampled function $\phi(xI/N, yI/N)$ with itself. The terms ε_τ and ε_T are Neumann factors taking the values $\varepsilon_0 = 1$ and $\varepsilon_\tau = 2$ for all $\tau \neq 0$. Equation (54) can be expressed as

$$\sigma^2_{F_{N,I}}(u, v) = \sum_{\tau,T=1}^{N-1}\sum^{N-1} \varepsilon_\tau \varepsilon_T \left(1 - \frac{\tau I}{N}\right)\left(1 - \frac{TI}{NI}\right)$$
$$\times \rho\left(\frac{\tau I}{N}, \frac{TI}{N}\right) \cos\left[\frac{2\pi I}{N}(u\tau + vT)\right] \quad (55)$$

This formulation is the Riemann approximating summation for large N for

$$S_I(u, v) = c \int_{-I}^{I}\int_{-I}^{I}\left(1 - \frac{|z_1|}{I}\right)\left(1 - \frac{|z_2|}{I}\right)$$
$$\times \rho(z_1, z_2) \cos[2\pi(uz_1 + vz_2)] \, dz_1 dz_2 \quad (56)$$

where c is a normalization constant and the continuous variables z_1 and z_2 have replaced the sampled variables $\tau I/N$ and TI/N, respectively. From Bochner's theorem it is known that

$$\rho(z_1, z_2) = R \int_{-\infty}^{\infty}\int_{-\infty}^{\infty} D(u, v) \exp\{-2\pi i(uz_1 + vz_2)\} \, du \, dv \quad (57)$$

where R is a constant chosen so that $D(u, v)$ has the form of a probability density function [38, p. 207]. Substitution of the covariance function into Eq. (56) then yields

$$S_I(u, v) = cR \int_{-\infty}^{\infty}\int_{-\infty}^{\infty} D(u - u', v - v') K_I(u', v') \, du' \, dv' \quad (58)$$

where K_I is the two-dimensional product Fejer kernel. It is known that $S_I(u, v)$ approaches $cRD(u, v)$ uniformly on compact sets as I approaches infinity [39, p. 2]. Consequently, it is reasonable to assume that the variance, $\sigma^2_{F_{N,I}}(u, v)$, behaves approximately as the power spectral density, $D(u, v)$, of the process, ϕ.

The results of this analysis have been obtained by first letting the sampling interval approach zero and then letting the window of observation grow. It is important to mention that if the relaxation or correlation radius of the covariance function, ρ, is small compared to the interval of observation, then it is reasonable to assume that the variance, $\sigma^2_{F_{N,I}}(u, v)$, is already close to the power spectral density without increasing the observation window. A similar result can be obtained for the discrete two-dimensional Fourier transform, $F(u, v)$, by scaling the window of observation to unity and noting that $f(x, y)$ is the sampled version of the continuous process, ϕ. In this case

$$\sigma^2_{F_N}(u, v) = \sum_{\tau,T=0}^{N-1}\sum \varepsilon_\tau \varepsilon_T \left(1 - \frac{\tau}{N}\right)\left(1 - \frac{T}{N}\right)$$
$$\times \rho\left(\frac{\tau}{N}, \frac{T}{N}\right) \cos\left[\frac{2\pi}{N}(u\tau + vT)\right] \tag{59}$$

The above stochastic model indicates that for an uncorrelated process, the spectrum tends to be flat, and the variance of the spectral components of the Fourier transform of $f(x, y)$ are fairly constant over a large range of frequencies. Conversely, if $f(x, y)$ is a highly correlated process, the variance of $F(u, v)$ tends to be large toward the low frequencies and falls off rapidly toward the higher frequencies. It will be assumed that the samples, $f(x, y)$, are identically distributed with variance V^2.

It is convenient to express Eq. (59) in an expanded form in order to investigate certain limiting conditions. Therefore,

$$\rho^2_{F_N}(u, v) = \rho(0, 0) + 2\sum_{\tau=1}^{N-1}\left(1 - \frac{\tau}{N}\right)\rho\left(\frac{\tau}{N}, 0\right)\cos\frac{2\pi u\tau}{N}$$
$$+ 2\sum_{T=1}^{N-1}\left(1 - \frac{T}{N}\right)\rho\left(0, \frac{T}{N}\right)\cos\frac{2\pi vT}{N}$$
$$+ 4\sum_{\tau,T=1}^{N-1}\sum\left(1 - \frac{\tau}{N}\right)\left(1 - \frac{T}{N}\right)$$
$$\times \rho\left(\frac{\tau}{N}, \frac{T}{N}\right)\cos\frac{2\pi}{N}(u\tau + vT) \tag{60}$$

For a random process, $f(x, y)$, which is constantly correlated in one direction, x, with correlation K, and totally uncorrelated in the other direction, the variance becomes

$$\sigma^2_{F_N}(u, v) = \rho(0, 0) - K + NK\delta(u) \tag{61}$$

and for the case where $\rho(0, 0) = K = V^2$, the variance of the identically distributed samples, $f(x, y)$, then

$$\sigma^2_{F_N}(u, v) = V^2 N \, \delta(u) \qquad (62)$$

Equation (61) indicates that for highly correlated processes in one-dimension the off axis variances are reduced by an amount equal to the one-dimensional correlation, K, and the variances on the axis corresponding to the correlated direction are increased by amount proportional to the correlation, K. For the case where the one-dimensional correlation equals the variance of the process, Eq. (62), all off-axis variances are zero and large variances are experienced on the correlation axis. For constant correlation, K, in both directions the variance behaves as

$$\sigma^2_{F_N}(u, v) = \rho(0, 0) - K + N^2 K \, \delta(u, v) \qquad (63)$$

and when the correlation equals the variance of the $f(x, y)$ process, the resulting frequency sample variance is

$$\sigma^2_{F_N}(u, v) = N^2 V^2 \, \delta(u, v) \qquad (64)$$

These results indicate that a process $f(x, y)$ with constant correlation equal to its variance in all directions is a deterministic constant with a Fourier transform equal to the Kronecker delta function at the origin.

Another limiting condition that is of interest is the case of total statistical independence of all samples in the process $f(x, y)$. In this case the variance of $F(u, v)$ becomes

$$\sigma^2_{F_N}(u, v) = V^2 \qquad (65)$$

This result indicates that for a statistically independent process all frequencies have identical variances. Under the condition of statistical independence of the samples, $f(x, y)$, the variance is sufficient to determine the distribution of frequency components. The central limit theorem applies assunming the $f(x, y)$ samples are bounded and identically distributed, and in the limit the distribution of the function $F(u, v)$ becomes normal [40, p. 294].

It is of interest to determine how closely to the normal the distributions of frequency samples behave for correlation in the process $f(x, y)$. Work has been done in this area in the one-dimensional case from the point of view of a strong mixing criterion for an ergodic process [41, p. 191]. Also, Diananda and others have proven theorems for limiting normal distributions for the r-dependent one-dimensional random process [42]. Expansion to the two-dimensional case is probable, but is not undertaken here.

6.2.2. Energy Distribution

If an original image $f(x, y)$ ranges in magnitude in units steps from 0 to A then the maximum magnitude of a transform sample will be AN and the minimum nonzero magnitude will be $1/N$. For example, the Hadamard transform of an image for which $f(x, y) = A$ is $F(u, v) = NA\,\delta(u, v)$. The Hadamard transform of $f(x, y) = 1 \cdot \delta(x, y)$ is $F(u, v) = 1/N$. Hence the dynamic range of transform samples in integer arithmetic is 1 to $N^2 A$.

While the dynamic range of variables in the transform domain is extremely large, it is interesting to note that only a few points can actually take on large values since the image energy in the spatial and transform domains is identical. This energy equivalence relationship can be derived from a generalization of Parseval's relationship [43] as follows. Let

$$F(u, v)F^*(u, v) = \sum_{x=0}^{N-1}\sum_{y=0}^{N-1} f(x, y)a(x, y, u, v)$$
$$\times \sum_{\alpha=0}^{N-1}\sum_{\beta=0}^{N-1} f(\alpha, \beta)a^*(\alpha, \beta, u, v) \qquad (66)$$

Expanding the product of the series yields

$$F(u, v)F^*(u, v) = \sum_{x=0}^{N-1}\sum_{y=0}^{N-1} [f(x, y)]^2 a(x, y, u, v)\, a^*(x, y, u, v)$$
$$+ \underbrace{\sum_{x=0}^{N-1}\sum_{y=0}^{N-1}\sum_{\alpha=0}^{N-1}\sum_{\beta=0}^{N-1}}_{x \neq \alpha,\, y \neq \beta} f(x, y)\, f(\alpha, \beta)\, a(x, y, u, v)\, a^*(\alpha, \beta, u, v) \qquad (67)$$

Now summing both sides over u and v gives

$$\sum_{u=0}^{N-1}\sum_{v=0}^{N-1} |F(u, v)|^2 = \sum_{x=0}^{N-1}\sum_{y=0}^{N-1} [f(x, y)]^2 \qquad (68)$$

as a result of the orthogonality property of the transform kernels as stated in Eq. (13). Hence, there is an energy equivalence between the spatial and transform domains.

6.3 CONCLUSIONS

The classic problem in the design of digital image coding systems is defined in the introduction to this chapter. A suggestion for the use of the transform domain of an image for possible coding in communication systems is made. Computer implemented Fourier and Hadamard transforms on various test scenes are presented as initial motivation for the study of transform domains for bandwidth reduction and error or noise immunity. The mathematics of image transformations are defined in terms of matrix theory and the question of separability and symmetric separability is considered. The mathematics of the Fourier transform in two dimensions is next discussed and certain properties of the transform domain are mentioned. The Hadamard transform is then described in two-dimensional notation and the concepts of sequency for image processing are presented. Brief mention of the Karhunen–Loeve transform for a class of images from a stochastic process is provided as background to a bandwidth reduction technique to be described in Chapter 7. Both statistical and energy analysis of the transform domain for the Fourier case are presented with mathematical rigor as a basis and foundation for the quantization procedures necessary for an equal bandwidth, also to be described in Chapter 7. With the elements of this chapter as background, we can now proceed to the study of the quantization of image transform domains and the implementation of bandwidth reduction techniques and noise immunity coding for improved digital image communication systems.

REFERENCES

1. W. K. Pratt, "A Bibliography on Television Bandwidth Reduction Studies," *IEEE Trans. Information Theory* **IT-13**, No. 1, 114–115 (January 1967).
2. A. Rosenfeld, "Bandwidth Reduction Bibliography," *IEEE Trans. Information Theory* **IT-14**, No. 4, 601–602 (July 1968).

3. *Proc. IEEE.* special issue on redundancy reduction, **55**, No. 3 (March 1967).
4. J. W. Cooley, and J. W. Tukey, "An Algorithm for the Machine Calculation of Complex Fourier Series," *Math. Computation* **19**, No. 90, 297-301 (1956).
5. W. T. Cochran, *et al.* "What is the Fast Fourier Transform?," *Proc. IEEE* **55**, No. 10 1664-1673 (October 1967).
6. J. W. Cooley, P. A. Lewis, and P. D. Welch, "Historical Notes on the Fast Fourier Transform," *Proc. IEEE* **55**, 1675-1677 (October 1967).
7. E. O. Brigham, and R. E. Morrow, "The Fast Fourier Transform," *IEEE Spectrum* **4**, No. 12, 63-70 (December 1967).
8. H. C. Andrews "A High Speed Algorithm for the Computer Generation of Fourier Transforms," *IEEE Trans. Computers* **C-17**, No. 4, 373 (April 1968).
9. H. C. Andrews, "Fourier Coding of Images," USCEE Rept No. 271, Unviersity of Southern California, Electronic Sciences Laboratory (June 1968).
10. H. C. Andrews and W. K. Pratt, "Fourier Transform Coding of Images," *Internat. Conf. System Sci. Hawaii 1968*, pp. 677-679 (January 1968).
11. H. C. Andrews and W. K. Pratt, "Televisision Bandwidth Reduction by Fourier Image Coding," *Soc. Motion Picture and Television Engs. 103rd Techn. Conf.* (May 1968).
12. H. C. Andrews and W. K. Pratt, "Television Bandwidth Reduction by Encoding Spatial Frequencies," *Soc. Motion Picture and Television Engrs.* **77**, 1279-1281 (December 1968).
13. W. K. Pratt, J. Kane, and H. C. Andrews, "Hadamard Transform Image Coding," *Proc. IEEE* **57**, No. 1 (January 1969).
14. H. C. Andrews and W. K. Pratt, "Transformation Coding for Noise Immunity and Bandwidth Reduction," *Second Annual Hawaii Internatl. Conf. System Sciences* (January 1969).
15. W. K. Pratt and H. C. Andrews, "Two-Dimensional Transform Coding of Images," *Internat. Symp. Information Theory* (January 1969).
16. J. Hadamard, "Resolution d'une Question Relative aux Determinants," *Bull. Sci. Math. Ser.* 2 **17**, Part I, 240-246 (1893).
17. H. J. Ryser, *Combinatorial Mathematics.* Wiley, New York 1963.
18. S. W. Golomb, *et al. Digital Communications.* Prentice-Hall, Englewood Cliffs, New Jersey, 1964.
19. R. E. Paley, "On Orthogonal Matrices," *J. Math. Phys.* **12**, 311-320 (1933).
20. J. Williamson, "Hadamard's Determinant Theorem and the Sum of Four Squares," *Duke Math. J.* **11**, 65-81 (1944).
21. L. Baumert, S. W. Golomb, and M. Hall, Jr., "Discovery of an Hadamard Matrix of Order 92," *Bull. Am. Math. Soc.* **68**, 237-238 (1962).
22. H. F. Harmuth, "A Generalized Concept of Frequency and Some Applications," *IEEE Trans. Information Theory* **IT-14**, No. 3, 375-382 (May 1968).
23. J. L. Walsh, "A Closed Set of Orthogonal Functions," *Am. J. Math.* **45**, 5-24 (1923).
24. N. J. Fine, "On the Walsh Functions," *Trans. Am. Math. Soc.* **65**, 373-414 (1949).

25. N. J. Fine, "The Generalized Walsh Functions," *Trans. Am. Math. Soc.* **69**, 66-72 (1950).
26. G. W. Morgenthaler, "On Walsh-Fourier Series," *Trans. Am. Math. Soc.* **84**, 472-507 (1957).
27. K. W. Henderson, "Some Notes on the Walsh Functions," *IEEE Trans. Electron. Computers* **EC-13**, 50-52 (February 1964).
28. H. Rademacher, "Einige Satze von Allgemeinen Orthogonal-Funktionen," *Ann. Math.* **87**, 122-138 (1922).
29. H. P. Dramer and M. V. Mathews, "A Linear Coding for Transmitting a Set of Correlated Signals," *IRE Trans. Information Theory* **IT-2**, 41-46 (September 1956).
30. J. E. Whelchel, Jr., and D. F. Guinn, "The Fast Fourier-Hadamard Transform and Its Use in Signal Representation and Classification," *Eascon 1968 Convention Record* pp. 561-573 (1968).
31. W. B. Davenport, Jr., and W. L. Root, *An Introduction to the Theory of Random Signals and Noise.* McGraw-Hill, New York, (1958).
32. J. J. Y. Huang and P. M. Schutheiss, "Block Quantization of Correlated Gaussian Random Variables," *IEEE Trans. Commun. Systems* **CS-11**, No. 3, 298-296 (September 1963).
33. T. Y. Young and W. H. Huggins, "On the Representation of Electrocardiographs," *IEEE Trans. Bio-Med. Electron.* **BME-10**, No. 3 86-95 (July 1963).
34. L. M. Goodman, "A Binary Linear Transformation for Redundancy Reduction," *Proc. IEEE Letters* **55**, No. 3, 467-468 (March 1967).
35. C. A. Andrews, J. M. Davies, and G. R. Schwarz, "Adaptive Data Compression," *Proc. IEEE* **55**, No. 3, 267-277 (March 1967).
36. C. J. Palermo, R. V. Palermo, and H. Horowitz, "The Use of Data Omission for Redundancy Removal," *Rec. Internat. Space Electron. Telemetry Sym. 1965*, pp. (11)D1-(11)D16 (1965).
37. W. M. Gentleman, "Matrix Multiplication and Fast Fourier Transformations," *Bell System Techn. J.* **47**, 1099-1103 (July-August 1968).
38. M. Loeve, *Probability Theory.* Van Nostrand, Princeton, New Jersey, 1955.
39. S. Bochner, *Harmonic Analysis and the Theory of Probability.* Univ. California Press, Berkeley, California, 1955.
40. B. V. Gnedenko, *The Theory of Probability.* Chelsea, New York, 1962.
41. Y. A. Rozanov, *Stationary Random Processes.* Holden-Day, San Francisco, (1967).
42. P. H. Diananda, "The Central Limit Theorem for m-Dependent Variables," *Proc. Cambridge Phil. Soc.* **51**, 92-95 (1955).
43. A. Papoulis, *The Fourier Integral and Its Applications.* McGraw-Hill, New York, 1962.

Chapter 7 IMAGE CODING[1]

7.0 INTRODUCTION

Chapter 6 laid the groundwork for use of the image transform domain as a means of coding for communication systems. It remains to show practical means of implementing those ideas. This chapter is devoted to the study of quantization techniques for equal bandwidth systems even before bandwidth reduction schemes can be considered. Various bandwidth reduction techniques are then implemented on the quantized transform domains and experimental results are presented. Noise immunity coding principles are then investigated in the transform domain and are shown to be fairly successful for high error rate binary symmetric channels.

To analyze the theoretical efficiency of coding the transform of a scene rather than the scene itself, it is necessary to compare the entropy of the spatial and transform domains. Andrews has shown that the entropy of a scene and its Fourier transform are identical [1]. The result holds true for any transform whose Jacobian is unity. This property of image transforms, though interesting, only establishes that under ideal coding the scene and its transform can be transmitted with the same channel capacity. It remains necessary to determine quantization and coding rules for practical channels.

---

[1] This chapter is by William Pratt and Harry Andrews.

136 IMAGE CODING

7.1 QUANTIZATION OF IMAGE TECHNIQUES

The selection of quantization levels can be made on the basis of minimizing the quantization error or achieving a uniform entropy for quantized sample amplitudes. In either case it is necessary to know the range and statistical distribution of the transform component to be quantized. Since this information is not available unless the transform is specified, quantization methods can only be investigated for particular transforms. Quantization methods for the Fourier and Hadamard transforms are considered in the following discussion.

Fourier transform samples are complex numbers which may be represented in real and imaginary, or magnitude and phase, form. In either case there are two components per transform sample that must be quantized. As a consequence of the statistical analysis of transform samples of Chapter 6, the real, $F_R(u, v)$, and imaginary, $F_I(u, v)$, components of the Fourier transform samples are assumed to follow the same Gaussian distribution whose variance, $\sigma^2(u, v)$, is proportional to the power spectral density of the original image. Hence

$$p\{F_R(u, v)\} = [2\pi\sigma^2(u, v)]^{-1/2} \exp\left\{-\frac{F_R^2(u, v)}{2\sigma^2(u, v)}\right\} \quad (1a)$$

$$p\{F_I(u, v)\} = [2\pi\sigma^2(u, v)]^{-1/2} \exp\left\{-\frac{F_I^2(u, v)}{2\sigma^2(u, v)}\right\} \quad (1b)$$

If the real and imaginary components are Gaussian, the magnitude of the Fourier transform sample, $F_M(u, v)$, is Rayleigh distributed:

$$p\{F_M(u, v)\} = \frac{F_M(u, v)}{\sigma^2(u, v)} \exp\left\{-\frac{F_M^2(u, v)}{2\sigma^2(u, v)}\right\}, \quad F_M(u, v) \geq 0 \quad (2a)$$

and its phase, $F_p(u, v)$, is uniformly distributed:

$$p\{F_p(u, v)\} = \frac{1}{2\pi}, \quad -\pi \leq F_p \leq +\pi \quad (2b)$$

Hadamard transform samples are real, bipolar numbers which can be represented by a single component per sample. For this analysis the statistical distribution of Hadamard sample components, $F_H(u, v)$, will be considered to follow a Gaussian distribution of the form,

$$p\{F_H(u, v)\} = [2\pi\sigma^2(u, v)]^{-1/2} \exp\left\{-\frac{F_H^2(u, v)}{2\sigma^2(u, v)}\right\} \quad (3)$$

When the variance function, $\sigma^2(u, v)$, is not known for a particular image, or class of images, to be transformed, the function can usually be modeled without seriously affecting the quantization process. From examination of the Fourier and Hadamard transforms of a typical image, it can be deduced that the variance function should be a maximum at the origin in the transform domain, be circularly symmetric, and decrease in magnitude monotonically toward the higher spatial frequencies. A two-dimensional function processing these characteristics is the Gaussian curve described by

$$\sigma^2(u, v) = S \exp\left\{-\frac{u^2 + v^2}{p}\right\} \qquad (4)$$

where S is an amplitude scaling constant and p is a spread control constant.

In the quantization analysis the transform sample component to be quantized (amplitude, real part, imaginary part, magnitude, or phase) is represented by the function $F_c(u, v)$. The range of the component is broken up into K positive and K negative bands separated by quantization levels Q_j ($j = 0, \pm 1, \pm 2, \ldots, \pm K$) where

$$Q_0 = 0 \qquad (5a)$$

$$Q_K = \frac{NA}{2} \qquad (5b)$$

$$Q_{-K} = -\frac{NA}{2} \qquad (5c)$$

The magnitude of a sample need only be quantized over the positive scale. If a transform component falls in a band bounded by quantization levels Q_{j-1} and Q_j, the component is quantized, and subsequently reconstructed, to the value F_j which lies within the band. The relationship between quantization levels and reconstruction levels is given below.

$$Q_{-K} = -\frac{NA}{2} \quad \underset{Q_{-K+1} \cdots Q_{-1}}{\overset{F_{-K} \quad F_{-2} \quad F_{-1} F_1 \quad F_{K-1} \quad F_K}{\big|}} \quad Q_0 \quad Q_1 \cdots Q_{K-1} \quad Q_K = +\frac{NA}{2}$$

reconstruction levels above, quantization levels below.

TABLE I

Quantization Error Criteria[a]

Cumulative mean square spatial error	$\sum_{x=0}^{N-1}\sum_{y=0}^{N-1} \overline{[f(x,y) - \tilde{f}(x,y)]^2}$
Cumulative mean square transform error	$\sum_{u=0}^{N-1}\sum_{v=0}^{N-1} \overline{[F(u,v) - \tilde{F}(u,v)]^2}$
Cumulative spatial error	$\sum_{x=0}^{N-1}\sum_{y=0}^{N-1} \lvert f(x,y) - \tilde{f}(x,y) \rvert$
Cumulative transform error	$\sum_{u=0}^{N-1}\sum_{v=0}^{N-1} \lvert F(u,v) - \tilde{F}(u,v) \rvert$
Relative spatial error	$\dfrac{\lvert f(x,y) - \tilde{f}(x,y) \rvert}{\lvert f(x,y) \rvert}$
Relative transform error	$\dfrac{\lvert F(u,v) - \tilde{F}(u,v) \rvert}{\lvert F(u,v) \rvert}$

[a] $F(u,v)$ = quantized value of $F(u,v)$; $\tilde{f}(x,y)$ = inverse transform of quantized value of $F(u,v)$.

Table I lists some error criteria that might be considered in the selection of transform quantization and reconstruction levels. In general, the error criteria chosen will depend upon the application of the reconstructed images; for example, whether the image is to be used for subjective viewing or photometric measurements.

For subjective viewing the relative spatial error for low brightness images provides an indication of image quality. This relative spatial error criterion is predicated by the fact that incremental brightness changes in the reconstructed image are much more noticeable if the brightness level is low than if it is high. Thus, to minimize the relative spatial error, the density of quantization levels in the spatial domain should be greater at the lower amplitude levels. But since the brightness of every point of a reconstructed image is a function of the amplitude of a single transform sample, then by the same reasoning, the density of quantization levels should be greater for low levels transform samples. From psychological tests, it is known that the human viewer is very sensitive to the location of high frequency brightness transitions, but relatively insensitive to their actual magni-

7.1 QUANTIZATION OF IMAGE TECHNIQUES **139**

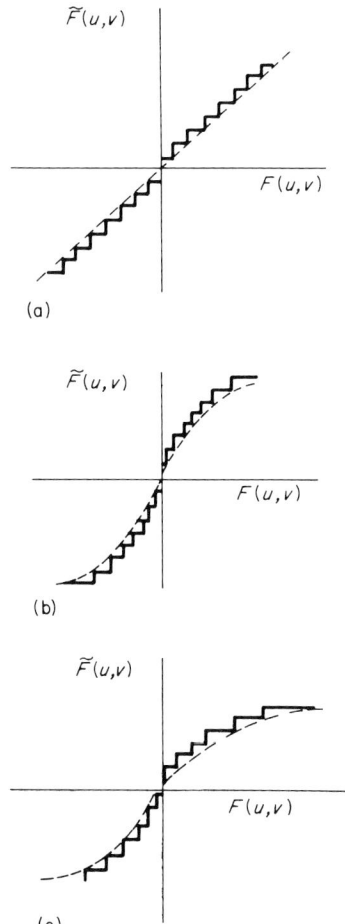

Fig. 1. Quantization rules. (a) Linear quantizer. (b) Gaussian quantizer. (c) Logarithmic quantizer.

140 IMAGE CODING

tude. In fact images which have been "crispened" by high pass filtering often appear preferable to the original image. From this characteristic of subjective viewing it would seem that the density of quantization levels at low transform sample amplitudes should be greater at the higher spatial frequencies than at the lower spatial frequencies. Thus, from the standpoint of subjective quality, the "best" quantizer should have a nonlinear characteristic such that the density of quantization levels over the range of the amplitude of the transform sample component to be quantized is:

    (a) greater at its lower values for a given spatial frequency
    (b) greater at the higher frequencies for a given amplitude

Figure 1 exhibits several quantization laws that are useful for quantization under a subjective viewing error criterion. The uniform or linear quantizer is commonly employed for quantization of the phase of Fourier transform samples. The Gaussian quantizers follow the mathematical function

$$\tilde{F}_c(u, v) = \operatorname{erf}\left[\frac{F(u, v)}{\sqrt{2}\, K(u, v)}\right] \quad (6)$$

between the transform component, $F_c(u, v)$, and the quantized transform component, $\tilde{F}_c(u, v)$, where

$$\operatorname{erf}(x) \equiv \frac{2}{\sqrt{\pi}} \int_0^x \exp\{-z^2\}\, dz \quad (7)$$

is the Gaussian error function and $K(u, v)$ is a two-dimensional positive function monotonically decreasing with u and v. The Gaussian quantizer has the desired property that the spacing of quantization levels is closer for the lower amplitudes of $F_c(u, v)$ at a given spatial frequency u, v and closer for the higher values of v and u at a given value of $F_c(u, v)$. Furthermore, if $K(u, v)$ is set equal to the standard deviation of the transform samples, $\sigma(u, v)$, then the probability that a transform sample will be quantized to a given reconstruction level will be the same for all quantization levels. This results in a uniform entropy for all reconstruction levels, and therefore, a constant word length code may be used for each quantized sample. The logarithmic quantizer follows the function

$$\tilde{F}_c(u, v) = \ln[K(u, v)(F_c(u, v) + 1)] \quad (8)$$

7.1 QUANTIZATION OF IMAGE TECHNIQUES

in the positive quadrant and the inverted and reversed version of the function in the negative quadrant. This function has the same general characteristics as the Gaussian quantizer, but does not produce an equal entropy for quantized samples.

If photometric measurements are to be made on an image the cumulative mean square spatial error is a common fidelity criterion. For such a situation the quantization levels in the *transform* domain must be selected to minimize the cumulative mean square error in the *spatial* domain. Let

$$\mathscr{E}_S \equiv \sum_{x=0}^{N-1} \sum_{y=0}^{N-1} \overline{[f(x, y) - \tilde{f}(x, y)]^2} \qquad (9)$$

represent the cumulative mean square spatial error where $\tilde{f}(x, y)$ is the image reconstruction from the quantized transform samples, $\tilde{F}(u, v)$. Then, in a matrix formulation

$$\begin{aligned}\mathscr{E}_S &= \sum_{x=0}^{N-1} \sum_{y=0}^{N-1} \overline{\{[B][F][B] - [B][\tilde{F}][B]\}^2} \\ &= \sum_{x=0}^{N-1} \sum_{y=0}^{N-1} [B]\overline{[F - \tilde{F}]^2}[B] \end{aligned} \qquad (10)$$

where $[B]$ represents the reverse transformation matrix.

Minimization of $\mathscr{E}_S$ in the spatial domain therefore can be accomplished by the minimization of the mean square error,

$$\mathscr{E}(u, v) \equiv [F - \tilde{F}]^2,$$

in the transform domain for all spatial frequencies.

In the case of the Fourier transform, the mean square error of each component of a transform sample must be minimized. The mean square error of a transform component may be written in the explicit form

$$\mathscr{E}(u, v) = \mathscr{E}_+(u, v) + \mathscr{E}_-(u, v) \qquad (11)$$

where

$$\mathscr{E}_+(u, v) = \sum_{j=1}^{K} \int_{Q_{j-1}}^{Q_j} (F_c - F_j)^2 p(F_c) \, dF_c$$

and

$$\mathscr{E}_-(u, v) = \sum_{j=-1}^{-K} \int_{Q_{j+1}}^{Q_j} (F_c - F_j)^2 p(F_c) \, dF_c$$

where $p(F_c)$ is the probability density of the transform sample component to be quantized. If $p(F_c)$ is a symmetrical probability density about $Q_0 = 0$, then $\mathscr{E}_+(u, v)$ equals $\mathscr{E}_-(v, v)$. Regardless of the form of $p(F_c)$ the quantization rule determined by the minimization of $\mathscr{E}_+(u, v)$ is the same as that determined from $\mathscr{E}_-(u, v)$ because of the symmetry of the quantization scale. Hence, only $\mathscr{E}_+(u, v)$ will be considered in the following analysis.

For a large number of quantization levels the probability density of the transform samples may be represented by a constant value, $p(F_j)$, over the quantization band. Hence,

$$\mathscr{E}_+(u, v) \approx \sum_{j=1}^{K} p(F_j) \int_{Q_{j-1}}^{Q_j} (F_c - F_j)^2 \, dF_c$$

$$= \frac{1}{3} \sum_{j=1}^{K} p(F_j)[(Q_j - F_j)^3 - (Q_{j-1} - F_j)^3] \quad (12)$$

The optimum placing of the reconstruction level F_j within the range Q_{j-1} to Q_j can be determined by minimization of $\mathscr{E}_+(u, v)$ with respect to F_j. Setting

$$\frac{d\mathscr{E}_+(u, v)}{dF_j} = \frac{1}{3} p(F_j)\{-3(Q_j - F_j)^2 + 3(Q_{j-1} - F_j)^2\} = 0 \quad (13)$$

yields

$$F_j = \frac{Q_j + Q_{j-1}}{2} \quad (14)$$

Therefore, the condition for minimizing $\mathscr{E}_+(u, v)$ in the range Q_{j-1} to Q_j is to place the reconstruction level F_j at the midpoint between each pair of quantization levels. The general relationship between reconstruction and quantization levels for the constraint of Eq. (14) is shown below.

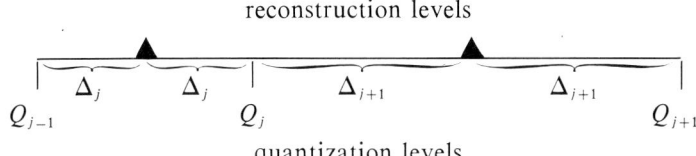

Hence,

$$Q_j = F_j + \Delta_j \tag{15}$$

and

$$Q_{j-1} = F_j - \Delta_j \tag{16}$$

where

$$\Delta_j \equiv \frac{Q_j - Q_{j-1}}{2} \tag{17}$$

With the value for F_j from Eq. (14) substituted into Eq. (12), the mean square transform error becomes

$$\mathscr{E}_+(u, v) \approx \frac{1}{12} \sum_{j=1}^{K} p[F_j][Q_j - Q_{j-1}]^3 \tag{18}$$

Now, by the definition of the integral of a function it is possible to write

$$\sum_{j=1}^{K} \{p[F_j]\}^{1/3}[Q_j - Q_{j-1}] = \int_0^{+NA/2} p^{1/3}[F_c] \, dF_c = \mathscr{K} \tag{19}$$

where the value of the integral, $\mathscr{K}$, is a constant only depending upon its limits. Thus, the problem of minimizing $\mathscr{E}_+(u, v)$ with respect to the reconstruction levels Q_j reduces to minimizing the sum of cubes of a variable in Eq. (18) subject to the constraint of Eq. (19) that the sum of the variables is constant. By the method of Lagrange multipliers $\mathscr{E}_+(u, v)$ is minimized when $\{p[F_j]\}^{1/3}[Q_j - Q_{j-1}]$ is identical for all K quantization bands. Under this condition

$$\{p[F_j]\}^{1/3}[Q_j - Q_{j-1}] = \frac{\mathscr{K}}{K} \tag{20}$$

and

$$[\mathscr{E}_+(u, v)]_{\min} = \frac{1}{12} \frac{\mathscr{K}^3}{K} = \frac{1}{12K^2} \left(\int_0^{+NA/2} \{p[F]^{1/3} \, dF \right)^3 \tag{21}$$

The quantization levels can be determined from the formula

$$Q_j = 2\Delta_1 + 2\Delta_2 + \cdots 2\Delta_{j-1} + \Delta_j \tag{22}$$

where from Eq. (20)

$$\Delta_j = \frac{\mathscr{K}}{2K\{p[F_j]\}^{1/3}} \tag{23}$$

Therefore,

$$Q_j = \frac{\mathscr{K}}{K}\left[\frac{1}{\{p[F_1]\}^{1/3}} + \frac{1}{\{p[F_2]\}^{1/3}} + \cdots + \frac{1}{\{p[F_{j-1}]\}^{1/3}}\right] \tag{24}$$

This series may be approximated by the normalized integral [2, 3]

$$Q_j = \frac{\frac{NA}{2}\int_0^{jNA/2K}\{p[F]\}^{-1/3}\,dF}{\int_0^{NA/2}\{p[F]\}^{-1/3}\,dF} \tag{25}$$

As an example of the computation of quantization levels by Eq. (25), consider the case for which $p(F) = 1/NA$. The quantization levels are then $Q_j = jNA/2K$ for $j = 1, 2, \ldots, K$.

If $p(F)$ is a Gaussian distribution with variance $\sigma^2(u, v)$ then Q_j is given by

$$Q_j = \frac{\frac{NA}{2}\int_0^{jNA/2K}\exp\left\{\frac{F^2}{6\sigma^2}\right\}dF}{\int_0^{NA/2}\exp\left\{\frac{F^2}{6\sigma^2}\right\}dF} \tag{26}$$

The quantization levels computed from Eq. (26) are more closely spaced for j small in the same general manner as the Gaussian or logarithmic quantizer.

Examples of image transform quantization for the Fourier and Hadamard transform using the uniform and Gaussian quantization laws are presented in the following paragraphs.

Reconstructions of the Fourier transform of the Surveyor spacecraft scenes with linear and Gaussian quantization are shown in Fig. 2. For both quantization rules 64 levels have been employed. The Gaussian quantization rule utilized a Gaussian shaped variance parameter with a spread control constant, $p = 500$. Results with the linear quantizer

7.1 QUANIZATION OF IMAGE TECHNIQUES **145**

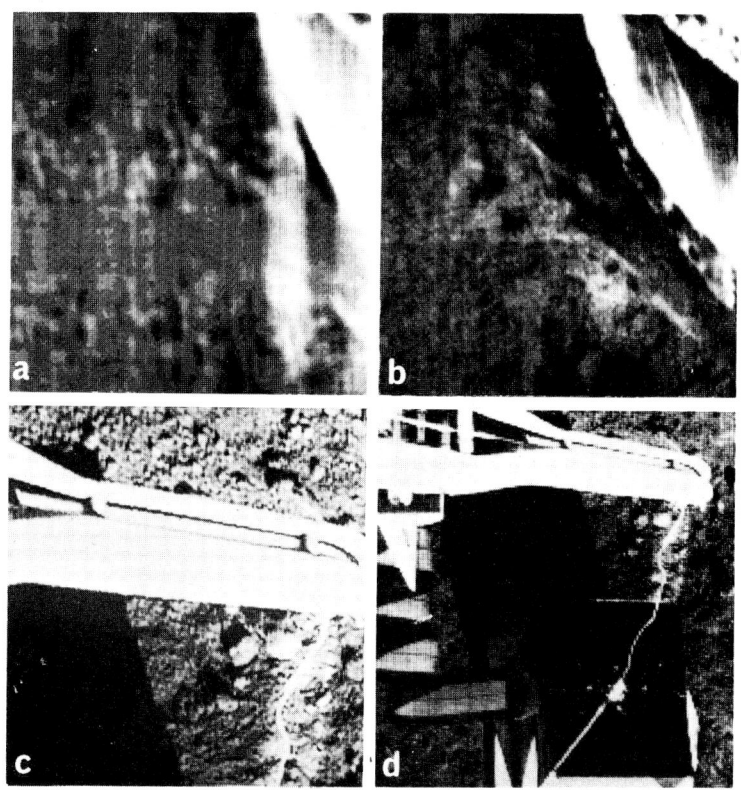

Fig. 2. 64-level quantization of Fourier transform. (a) Inverse Fourier transform of linearly quantized Fourier transform of footpad. (b) Inverse Fourier transform of Gaussian quantized Fourier transform of footpad, $p = 500$. (c) Inverse Fourier transform of Gaussian-quantized Fourier transform of boom, $p = 500$. (d) Inverse Fourier transform of Gaussian-quantized Fourier transform of box, $p = 500$.

146 IMAGE CODING

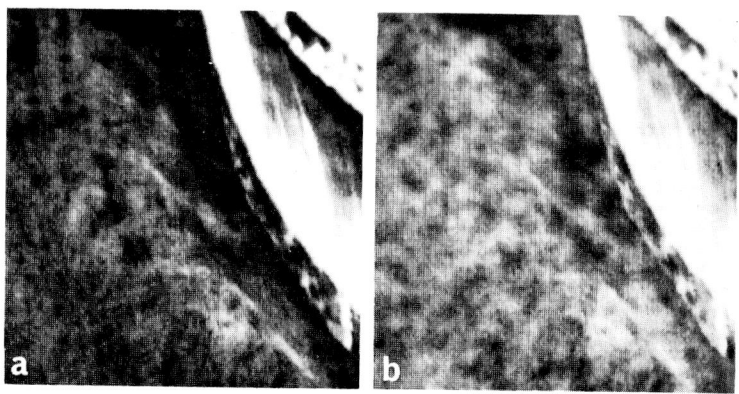

Fig. 3. 32- and 16-level Gaussian quanitzation of Fourier transform of footpad, $p = 500$. (a) Inverse Fourier transform of Gaussian-quantized Fourier transform, 32 levels. (b) Inverse Fourier transform of Gaussian-quantized Fourier transform, 16 levels.

for the footpad scene are poor because of the large quantization errors at high spatial frequencies. The Gaussian quantizer reconstructions for the footpad, boom, and box scenes show negligible image degradation.

Figure 3 illustrates tests to determine the effect of few quantization levels with the Gaussian quantizer. In these tests, reconstructions have been made with the Fourier transform samples quantized to 32 and 16 levels with a Gaussian shaped variance function with $p = 500$. The loss of resolution in these pictures is due to the quantization errors at high spatial frequencies.

If the Gaussian quantization rule is to be practical it is imperative that a variance function can be chosen for a wide class of scenes without detailed knowledge of the content of these scenes. Figure 2 shows that the Gaussian shaped variance function with the spread control parameter, $p = 500$, provides satisfactory reconstructions for three different scenes. It is also of interest to determine the effect of changes in the variance function. Figure 4 contains reconstructions using the Gaussian quantizer with 64 quantization levels for a Gaussian shaped variance function with a spread parameter of $p = 250$ and 1000, and with a $|(\text{sinc } au)(\text{sinc } av)|$ shaped variance function. These experiments indicate that fortunately the performance of the Gaussian

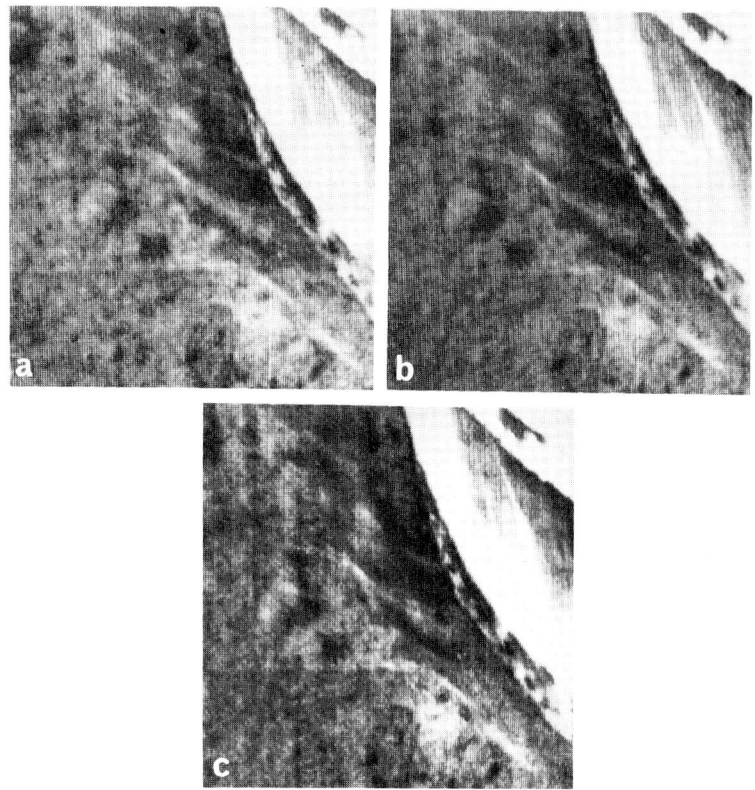

Fig. 4. 64-level Gaussian quantization of Fourier transform of footpad with different variance functions. (a) Inverse Fourier transform of Gaussian-quantized Fourier transform with a narrow Gaussian-shaped variance function, $p = 250$. (b) Inverse Fourier transform of Gaussian-quantized Fourier transform with a wide Gaussian-shaped variance function, $p = 1000$. (c) Inverse Fourier transform of Gaussian-quantized Fourier transform with a $|(\text{sinc } au)(\text{sinc } av)|$-shaped variance function.

148 IMAGE CODING

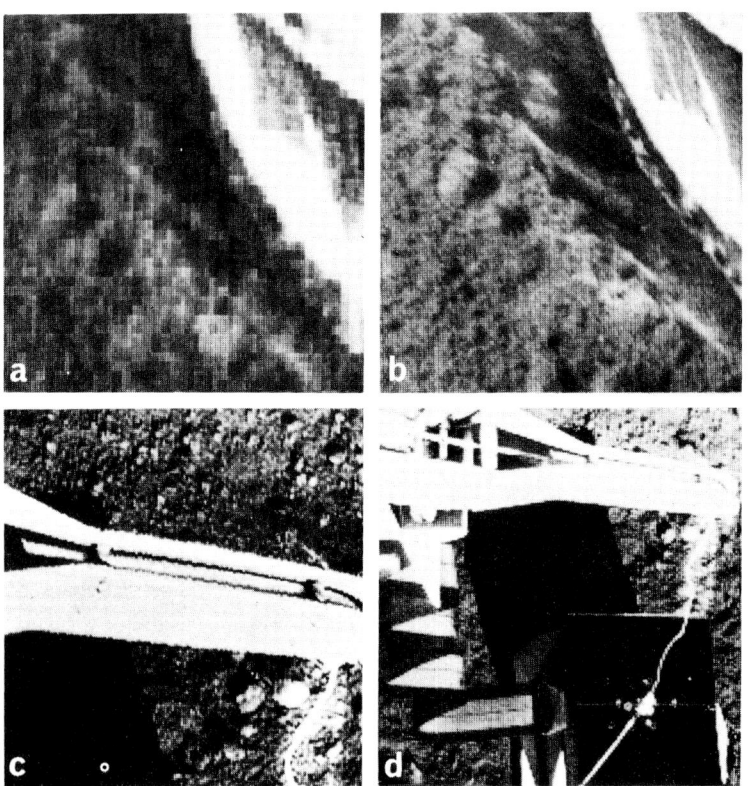

Fig. 5. 64-level quantization of Hadamard transform. (a) Hadamard transform of linearly quantized Hadamard transform of footpad. (b) Hadamard transform of Gaussian-quantized Hadamard transform of footpad, $p = 1500$. (c) Hadamard transform of Gaussian-quantized Hadamard transform of boom, $p = 1500$. (d) Hadamard transform of Gaussian-quantized Hadamard transform of box, $p = 1500$.

7.1 QUANTIZATION OF IMAGE TECHNIQUES **149**

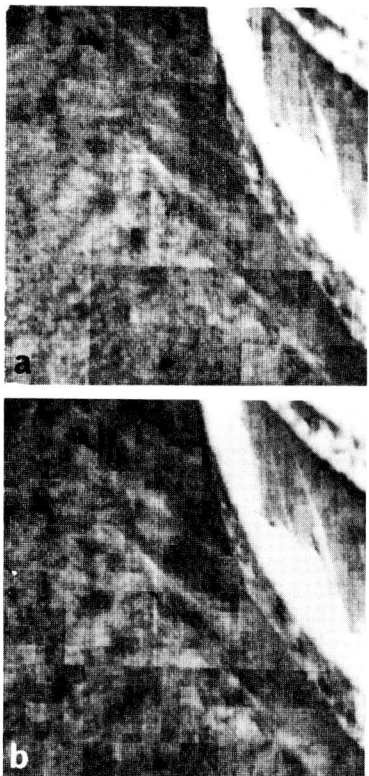

Fig. 6. 32- and 16-level Gaussian quantization of Hadamard transform of footpad, $p = 1500$. (a) Hadamard transform of Gaussian-quantized Hadamard transform, 32 levels. (b) Hadamard transform of Gaussian-quantized Hadamard transform, 16 levels.

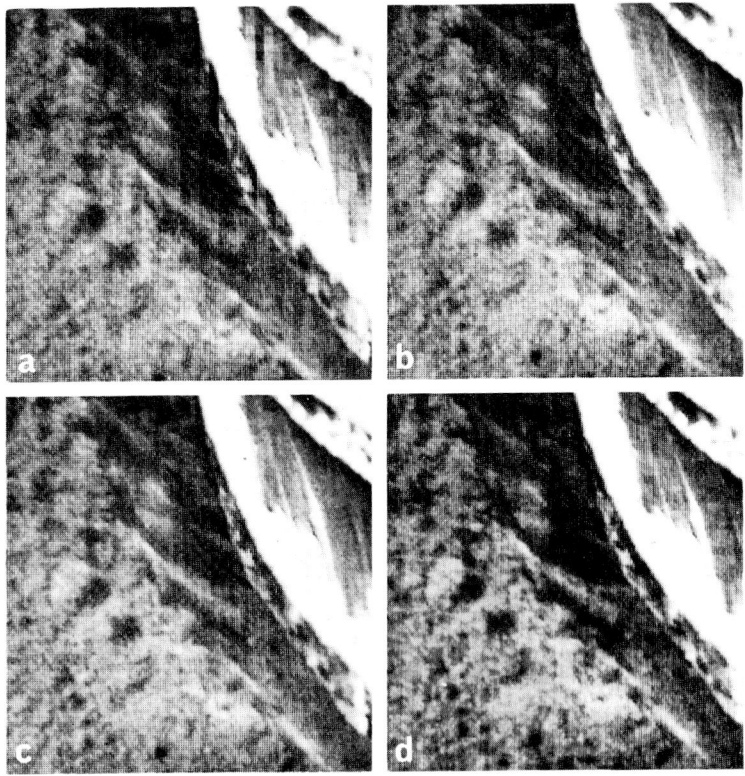

Fig. 7. 64-level Gaussian quantization of Hadamard transform of footpad with different Gaussian-shaped variance functions. (a) Hadamard transform of Gaussian-quantized Hadamard transform, $p = 500$. (b) Hadamard transform of Gaussian-quantized Hadamard transform, $p = 1000$. (c) Hadamard transform of Gaussian-quantized Hadamard transform, $p = 2000$. (d) Hadamard transform of Gaussian-quantized Hadamard transform, $p = 5000$.

quantizer is relatively insensitive to the exact mathematical form of the variance function.

The same set of quantization experiments has been performed for quantization of Hadamard transform samples with essentially the same results and conclusions. The results of these experiments shown in Figs. 5–7 are self explanatory.

The conclusions of the quantization experiments for the Fourier and Hadamard transform is that good quality reconstructions are possible when the transform samples have been quantized to as few as 64 levels using the Gaussian quantization rule and a Gaussian shaped variance function with an appropriate spread control parameter.

7.2 BANDWIDTH REDUCTION

Transmission of the transform of an image rather than the image itself opens up a wide area of investigation for the development of image transform bandwidth reduction techniques. Such techniques may be divided into two categories: those which are based upon the unique structure of the energy distribution in the transform plane, and those that seek to apply conventional spatial domain bandwidth reduction methods to the transform domain in a manner rather independent of its energy distribution. In general, the former class of methods provide the best performance. Attempts to apply spatial domain bandwidth reduction techniques to transform samples have not prove successful because of the large dynamic ranges of transform samples and their relative lack of correlation with one another.

7.2.1 Transform Sampling

Many transform bandwidth reduction techniques can be analyzed from the viewpoint of two-dimensional sampling. Figure 8 illustrates a generalized block diagram of a transform sampling system. The forward transform of an image, $F(u, v)$, is multiplied by a two-dimensional sampling function, $S(u, v)$, which takes on the values 0 or 1 according to some *a priori* or adaptive rule. Several transform sampling methods are listed in Table II. These methods will be considered individually in subsequent subsections.

152 IMAGE CODING

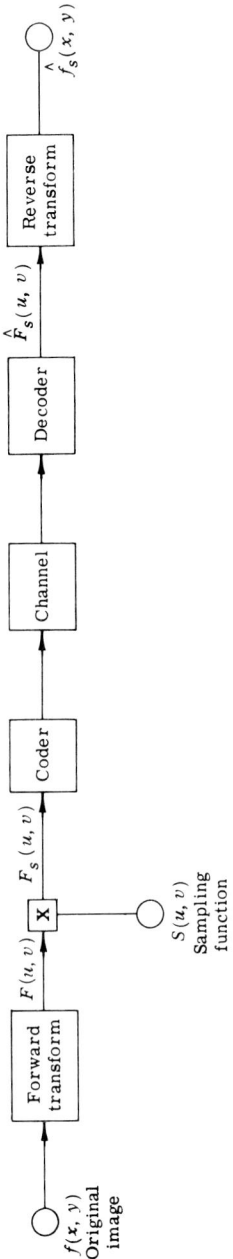

Fig. 8. Transform domain sampling.

TABLE II
Classifications of Transform Sampling Methods

Description	Sampling function $S(u, v)$	Conditions
Checkerboard sampling	$\dfrac{1 + (-1)^{u+v}}{2}$	Odd samples set to zero
Random sampling	1	Probability p
	0	Probability $1 - p$
Zonal sampling	1	$u, v \in$ sampling region
	0	$u, v \notin$ sampling region
Threshold sampling	1	If $|F(u, v)| > M_T(u, v)$
	0	If $|F(u, v)| \leq M_T(u, v)$

With reference to Fig. 8, the sampled transform, $F_s(u, v)$, is simply

$$F_s(u, v) = F(u, v)S(u, v) \qquad (27)$$

After decoding $\hat{F}_s(u, v)$ is reconstructed and the reverse transform produces $\hat{f}_s(x, y)$. The effect of channel errors on reduced bandwidth signals is an important topic which will be considered in the next section. For this analysis the transmission will be assumed to be errorless. With this assumption the factor of greatest importances becomes the closeness with which the reverse transform of $F_s(u, v)$ approximates the original image, $f_s(x, y)$. Taking the reverse transform of $F_s(u, v)$ yields

$$f_s(x, y) = \sum_{u=0}^{N-1} \sum_{v=0}^{N-1} F(u, v)S(u, v)b(x, y, u, v) \qquad (28)$$

Since

$$F(u, v) = \sum_{\alpha=0}^{N-1} \sum_{\beta=0}^{N-1} f(\alpha, \beta)a(\alpha, \beta, u, v) \qquad (29)$$

the system output image may be expressed as

$$f_s(x, y) = \sum_{u=0}^{N-1} \sum_{v=0}^{N-1} S(u, v)b(x, y, u, v) \sum_{\alpha=0}^{N-1} \sum_{\beta=0}^{N-1} f(\alpha, \beta)a(\alpha, \beta, u, v) \qquad (30)$$

Upon changing the order of summation

$$f_s(x, y) = \sum_{\alpha=0}^{N-1} \sum_{\beta=0}^{N-1} f(\alpha, \beta) \sum_{u=0}^{N-1} \sum_{v=0}^{N-1} S(u, v) a(\alpha, \beta, u, v) b(x, y, u, v) \quad (31)$$

For the Fourier case

$$a(\alpha, \beta, u, v) b(x, y, u, v) = \frac{1}{N} b(x - \alpha, y - \beta, u, v) \quad (32)$$

The second summation is then recognized to be the reverse transform of $s(u, v)$ evaluated at the point $\alpha - x$, $\beta - y$ in the spatial domain. Hence,

$$f_s(x, y) = \frac{1}{N} \sum_{\alpha=0}^{N-1} \sum_{\beta=0}^{N-1} f(\alpha, \beta) s(x - \alpha, y - \beta) \equiv f(x, y) \circledast s(x, y) \quad (33)$$

represents the spatial convolution denoted by the symbol $\circledast$, of the original image, $f(x, y)$, with the reverse transform of the sampling function, $S(u, v)$ when using the Fourier transform.

In general, any sampling function can be expressed as

$$S(u, v) = \frac{1 + R(u, v)}{2} \quad (34)$$

where $R(u, v)$ takes on the value ± 1 as a function of the spatial frequencies u and v. The reverse transform of the sampled transform domain is then

$$f_s(x, y) = \frac{1}{2} [f(x, y) + f(x, y) \circledast r(x, y)] \quad (35)$$

where $r(x, y)$ is the reverse transform of $R(u, v)$. Thus, the reconstruction of the sampled image is composed of the original image plus some additive interference that is dependent upon the form of the original image and the sampling function.

As an example of deterministic sampling, consider a sampling function

$$S(u, v) = \frac{1 + (-1)^{u+v}}{2} \quad (36)$$

which samples the Fourier transform of an image in a checkerboard pattern. For this case

$$R(u, v) = (-1)^{u+v} = \exp\{i\pi(u + v)\} \quad (37)$$

and its inverse Fourier transform is

$$r(x, y) = \sum_{u=0}^{N-1} \sum_{v=0}^{N-1} \exp\{i\pi(u + v)\} \exp\left\{\frac{-2\pi i}{N}(ux + vy)\right\} \quad (38)$$

or

$$r(x, y) = \delta\left(x + \frac{N}{2}, y + \frac{N}{2}\right) \quad (39)$$

Hence, the reconstructed image

$$f_s(x, y) = \frac{1}{2}\left[f(x, y) + f\left(x + \frac{N}{2}, y + \frac{N}{2}\right)\right] \quad (40)$$

is composed of the original image overlaid by the original shifted horizontally and vertically by one-half its size. Figure 9(a) illustrates the experimental verification of this effect for the footpad scene.

A nondeterministic sampling procedure that has been considered is one in which $R(u, v)$ is a random variable assuming the values ± 1. If this random variable is highly uncorrelated, the additive interference is spread out over the reconstructed image. This technique has been investigated for random sampling of the Fourier transform of an image in which 50 per cent of the transform components have been sampled at random positions to yield a bandwidth reduction of 2:1. The reconstructed image of the footpad scene is shown in Fig. 9(b). With this type of random sampling the convolutional interference produces a significant amount of image degradation. Distortion in this image is due principally to the convolution of the high brightness, and low spatial frequency, portion of the footpad over the image surface. To overcome this difficulty only the highest 90 percent of the spatial frequencies of the image were randomly sampled. The low spatial frequencies were completely sampled. The reconstruction in Fig. 9(c) for this type of sampling shows some improvement, but the image distortion is still severe.

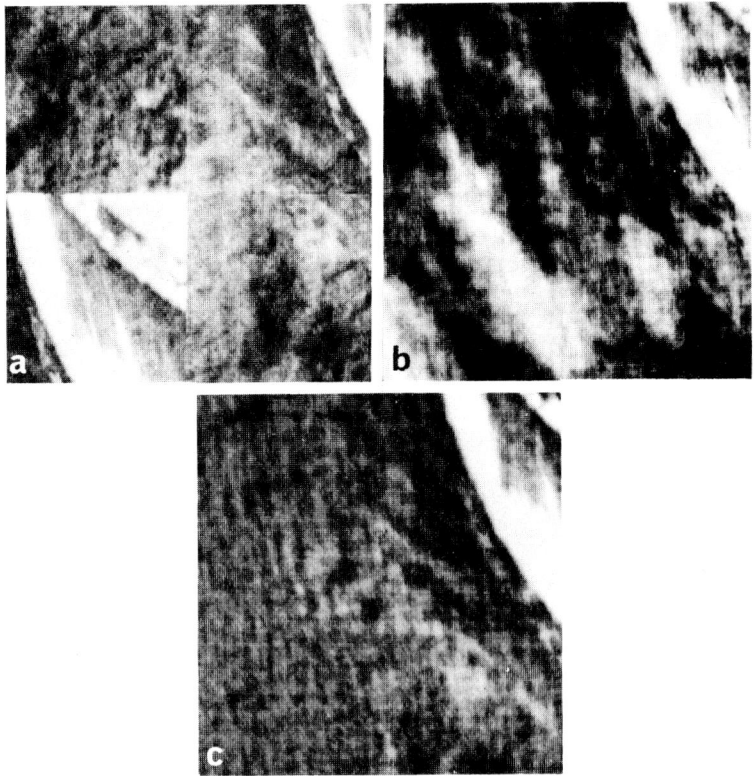

Fig. 9. Checkerboard and random sampling of Fourier transform. (a) Checkerboard sampling. (b) Random sampling. (c) Random sampling of high spatial frequencies only.

7.2.2 Zonal Sampling

In most scenes of interest there is a fairly high degree of correlation between adjacent image elements. For these types of images the energy in the transform plane tends to be clustered at certain spatial frequencies.

Figure 10 illustrates the percentage of energy within a circle centered at the origin of the Fourier transform plane for the three Surveyor spacecraft scenes. For all three scenes 95 percent of the image energy is contained in 1 percent or less of the Fourier domain samples. With an image energy distribution such as that shown in Fig. 10, the most obvious method of conserving bandwidth is simply to not transmit the high spatial frequency information. Discarding the high spatial frequencies is equivalent to passing the image through a circular, zonal low pass filter; the result is a loss of focus. Figures 11 and 12 show the effect of zonal low pass sampling of the Surveyor spacecraft footpad and box scenes. These experiments support the widely known fact that the high frequency brightness transitions are important even though they are relatively few in number and contain a low proportion of the image energy. However if some degree of resolution loss is

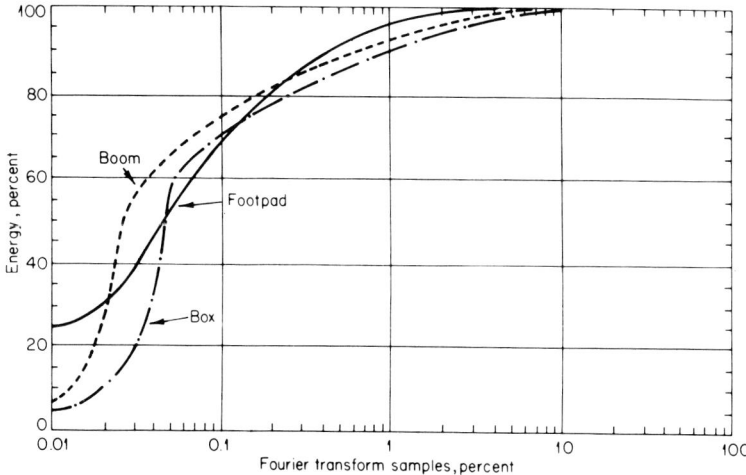

Fig. 10. Image energy as a function of percentage of Fourier transform samples within a Circular zone.

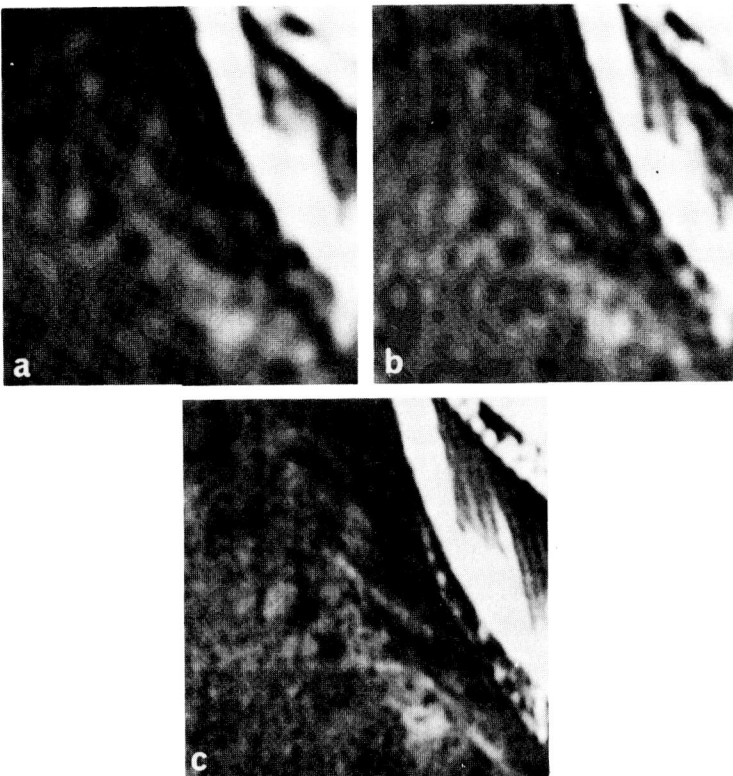

Fig. 11. Low pass zonal Fourier transform sampling, footpad. (a) 99.8% energy transmitted, 32:1 BWR. (b) 99.9% energy transmitted, 8:1 BWR. (c) 99.99% energy transmitted, 4:1 BWR.

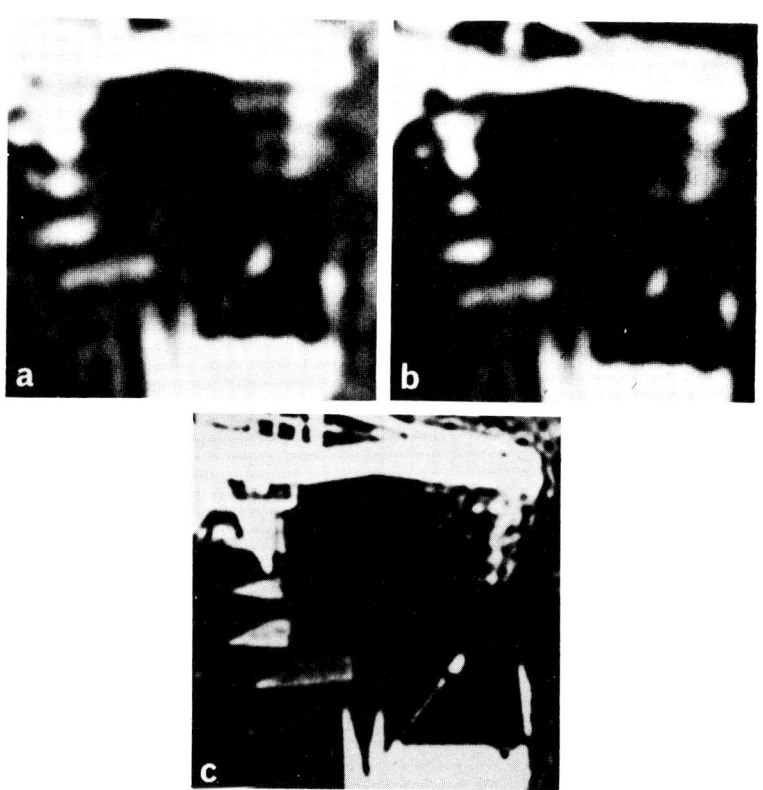

Fig. 12. Low pass zonal Fourier transform sampling, box. (a) 98.3% energy transmitted, 32:1 BWR. (b) 99.8% energy transmitted, 8:1 BWR. (c) 99.9% energy transmitted, 4:1 BWR.

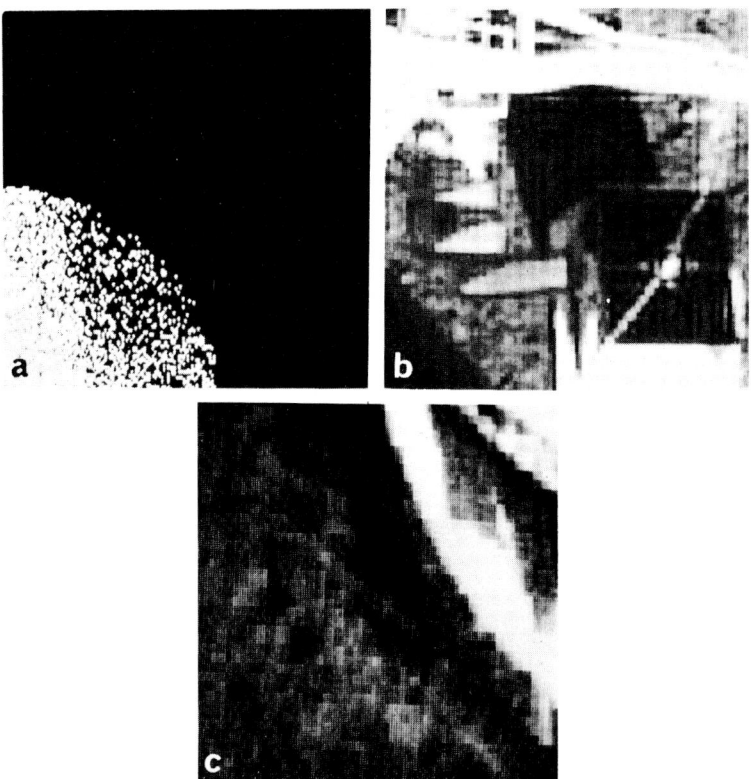

Fig. 13. Low pass zonal Hadamard transform sampling. (a) Threshold display of zonal filtered Hadamard transform. (b) 4:1 Sample reduction, box. (c) 4:1 Sample reduction, footpad.

acceptable, zonal low pass filtering of the Fourier domain does yield relatively large bandwidth reductions.

Zonal low pass sampling or filtering can also be performed in the Hadamard transform domain. Figure 13 illustrates a reconstruction of the zonal low pass Hadamard domain spatial sequencies of the footpad and box scenes. The image degradation tends to be more noticeable for zonal filtering of the Hadamard transform than for the Fourier transform for the same bandwidth reduction factor because of the rectangular shape of the two-dimensional Hadamard reconstruction waveforms. The eye is very sensitive to the presence of sharp brightness transitions within an image. With the Hadamard transform all transitions occur within one element, whereas in the Fourier transform the brightness transitions are spread over many elements since the reconstruction waveforms are two-dimensional sinusoids.

7.2.3 Threshold Sampling

The difficulty with the zonal filter sampling method of bandwidth reduction is that large magnitude samples are indiscriminately discarded. An obvious answer to this problem is to code only those samples whose magnitudes are above a given threshold level. With this coding method it becomes necessary to provide information as to the location of significant samples.

Selection of a threshold level for a given transform is generally a compromise between the number of samples deleted and the error resulting from the deletion of samples. If a constant threshold is chosen, then the maximum magnitude of deleted samples will be independent of spatial frequency, but the probability of deleting a given sample will usually be a function of its spatial frequency. On the other hand, if the threshold is chosen to be linearly dependent upon the variance of samples, the deletion probability in most cases will be constant for all samples, but the deletion error will be a function of spatial frequency. The "best" threshold can only be determined analytically for a given error criterion.

For transform threshold coding the "best" transform is one which maximizes the number of transform samples which are zero or near zero when the error criterion is satisfied. It has been pointed out in Chapter 6 that to minimize the mean square error between an original signal and a transform reconstruction for threshold sample deletion, the optimum transform is composed of eigenvectors of the correlation matrix of the data. Since mean square error has not proven an effective measure of error between images, the usefulness of this transform remains in doubt. Further investigation of the application of the eigenvector transform to image coding is required.

For a given transform, the expected bandwidth reduction achievable with threshold sampling of the transform domain can be estimated if the probability distribution of the magnitude of transform samples is known. Let $M(u, v) \equiv |F(u, v)|$ be the magnitude of a transform sample and $P[M(u, v)]$ be the probability distribution of the magnitude of transform samples, then the probability, $P(u, v)$, that the magnitude of a transform sample of spatial frequency (u, v) is greater than a threshold level, $M_T(u, v)$, is

$$P(u, v) = \int_{M_T(u, v)}^{\infty} P[M(u, v)] \, dM \tag{41}$$

The expected number of samples above the threshold, N_T, is then given by

$$N_T = \sum_u \sum_v P(u, v) \tag{42}$$

where the limits of the summation are dependent upon the type of transform employed.

Consider first Fourier threshold sampling. If it is assumed that the probability density of the real and imaginary components of transform samples are Gaussian with a variance function $\sigma^2(u, v)$, then the magnitude of a transform sample becomes Rayleigh distributed with a distribution

$$P[M(u, v)] = \frac{M}{\sigma^2(u, v)} \exp\left\{-\frac{M^2}{2\sigma^2(u, v)}\right\}, \quad M \geq 0 \tag{43}$$

Then the probability that the sample magnitude is greater than the threshold, $M_T(u, v)$, is

$$P(u, v) = \exp\left\{-\frac{[M_T(u, v)]^2}{2\sigma^2(u, v)}\right\} \tag{44}$$

7.2 BANDWIDTH REDUCTION

As a result of the conjugate symmetry property of the Fourier transform, only one half of the transform samples need be considered. The expected number of these samples above the threshold is given by

$$N_T = \sum_{u=0}^{N-1} \sum_{v=0}^{N-1} \exp\left\{-\frac{[M_T(u, v)]^2}{2\sigma^2(u, v)}\right\} \quad (45)$$

In the special case for which the threshold is linearly dependent upon the variance, i.e.,

$$M_T(u, v) = k_T \sigma(u, v) \quad (46)$$

where k_T is a constant, the expected number of samples above the threshold is simply

$$N_T = \frac{N^2}{2} \exp\left\{-\frac{k_T^2}{2}\right\} \quad (47)$$

The transform sample reduction is then

$$\frac{N^2/2}{N_T} = \exp\left\{\frac{k_T^2}{2}\right\} \quad (48)$$

For the Hadamard transform, the magnitude of the samples can be modeled as a Gaussian distribution.

$$P[M(u, v)] = [2\pi\sigma^2(u, v)]^{-1/2} \exp\left\{-\frac{M^2}{2\sigma^2(u, v)}\right\} \quad (49)$$

Then the probability that a transform sample exceeds the threshold becomes

$$P(u, v) = \left[1 - \text{erf}\left(\frac{M_T(u, v)}{\sqrt{2}\,\sigma(u, v)}\right)\right] \quad (50)$$

where

$$\text{erf}\{x\} \equiv \int_0^x e^{-y^2}\, dy \quad (51)$$

is the Gaussian error function. The expected number of transform samples above the threshold is

$$N_T = \sum_{u=0}^{N-1} \sum_{v=0}^{N-1} \left[1 - \text{erf}\left(\frac{M_T(u, v)}{\sqrt{2}\,\sigma(u, v)}\right)\right] \quad (52)$$

And for the special case for which $M_T(u, v) = k_T \sigma(u, v)$, the transform sample reduction becomes

$$\frac{N^2}{N_T} = \frac{1}{\left[1 - \mathrm{erf}\left(\frac{k_T}{\sqrt{2}}\right)\right]} \quad (53)$$

From the preceding analysis, it is seen that if the variance factor $\sigma^2(u, v)$ is known for a particular class of images, the expected bandwidth reduction factor can be easily computed for Fourier and Hadamard threshold sampling with a given threshold level. The amount of image degradation for a given threshold level must be determined at present by a subjective evaluation or comparative measurements.

Transform threshold coding experiments have been performed for the Fourier and Hadamard transforms. Figures 14 and 15 show the location of samples above a threshold and the corresponding reconstructions from these samples for the Fourier transform of the boom scene. If the threshold becomes too high, the loss of the high spatial frequency samples becomes noticeable. For this scene a threshold level of 200 provides good quality reconstructions. In this particular scene the magnitude of the largest, nonzero, spatial frequency is 53,186. The number of Fourier domain samples in the half plane above the threshold, and the equivalent sample reduction are listed (Table III)

TABLE III

Threshold level	Number of fourier domain samples above threshold	Sample reduction
100	13,402	2.4
200	4,887	6.2
300	2,532	12.9

Figure 16 illustrates the results of the Fourier transform threshold coding experiment for the box scene. Again, a threshold level of 200 provides a good reconstruction.

Similar results have been obtained for threshold coding of Hadamard transform samples. Figure 17 compares the Hadamard threshold coding for the box scene.

7.2 BANDWIDTH REDUCTION 165

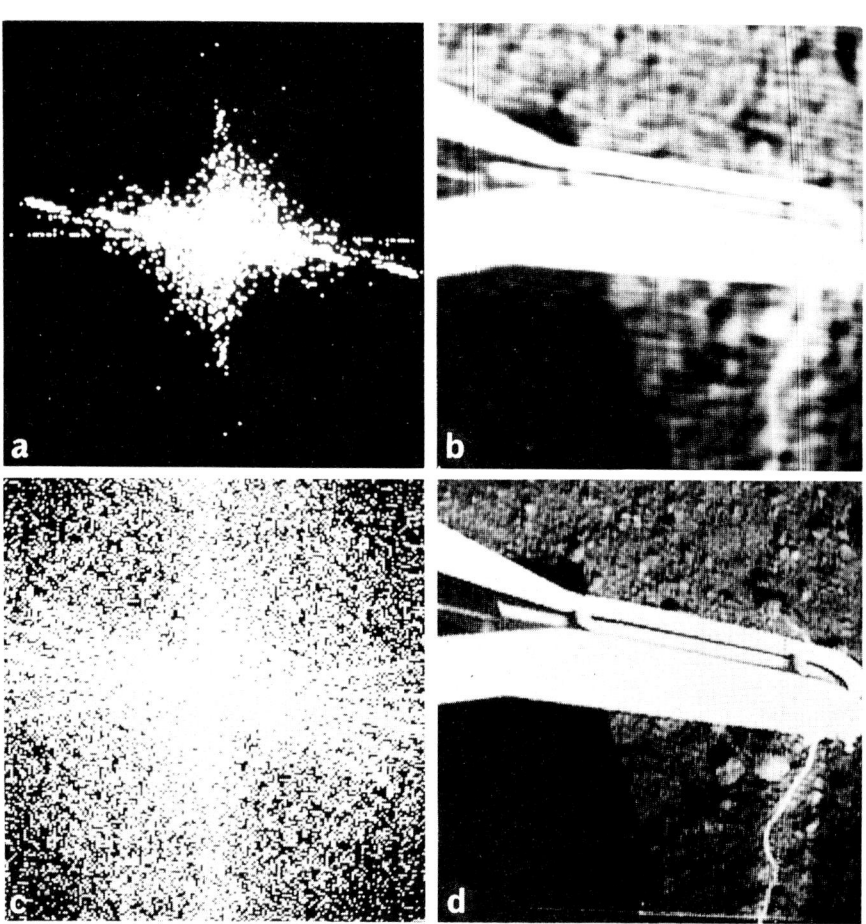

Fig. 14. Fourier transform threshold sampling with high and low thresholds, boom. (a) Map of samples above 500. (b) Fourier transform of 500-level thresholded samples. (c) Map of samples above 100. (d) Fourier transform of 100-level thresholded samples.

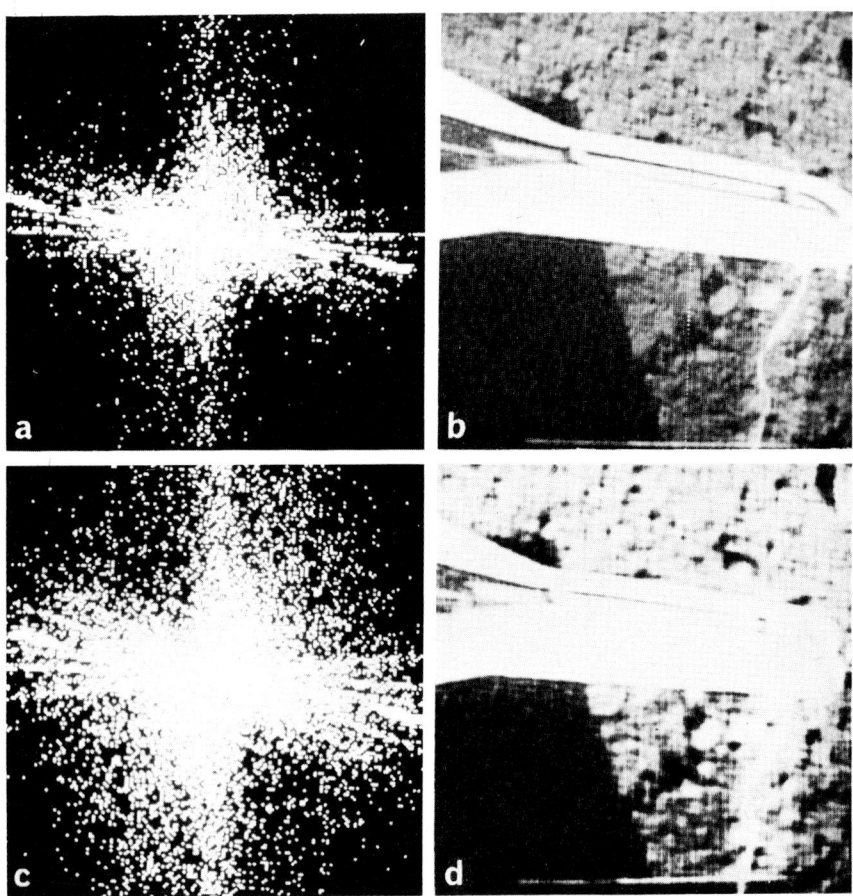

Fig. 15. Fourier transform threshold sampling with intermediate thresholds, boom. (a) Map of samples above 300. (b) Fourier transform of 300-level thresholded samples. (c) Map of samples above 200. (d) Fourier transform of 200-level thresholded samples.

7.2 BANDWIDTH REDUCTION 167

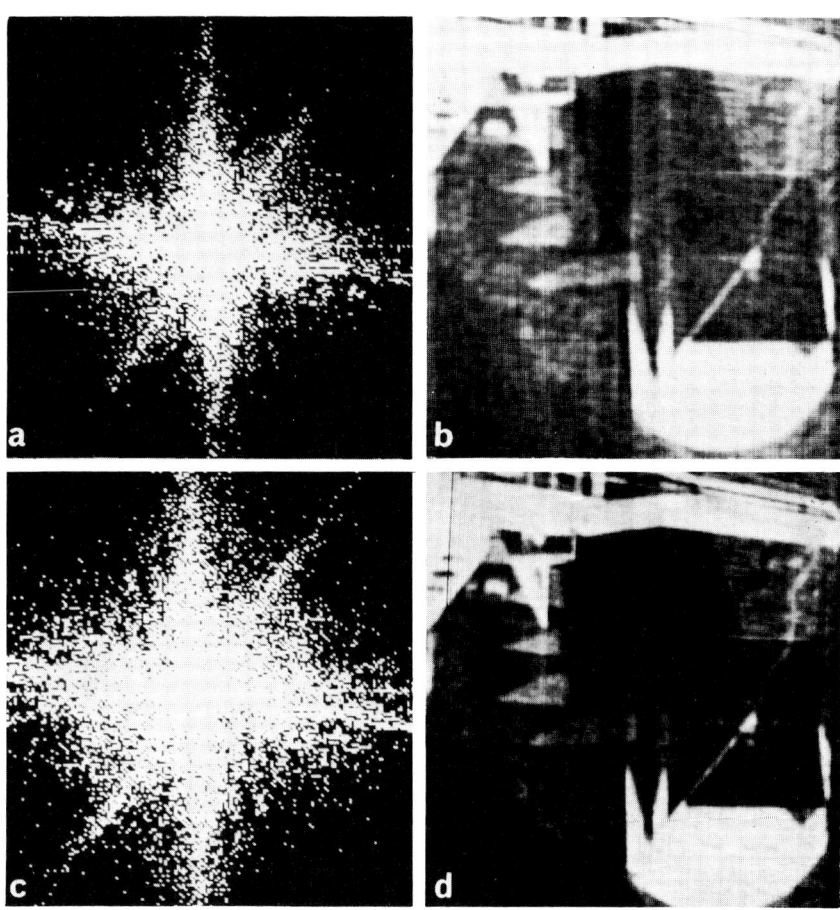

Fig. 16. Fourier transform threshold sampling with intermediate thresholds, box. (a) Map of samples above 300. (b) Fourier transform of 300-level thresholded samples. (c) Map of samples above 200. (d) Fourier transform of 200-level thresholded samples.

168 IMAGE CODING

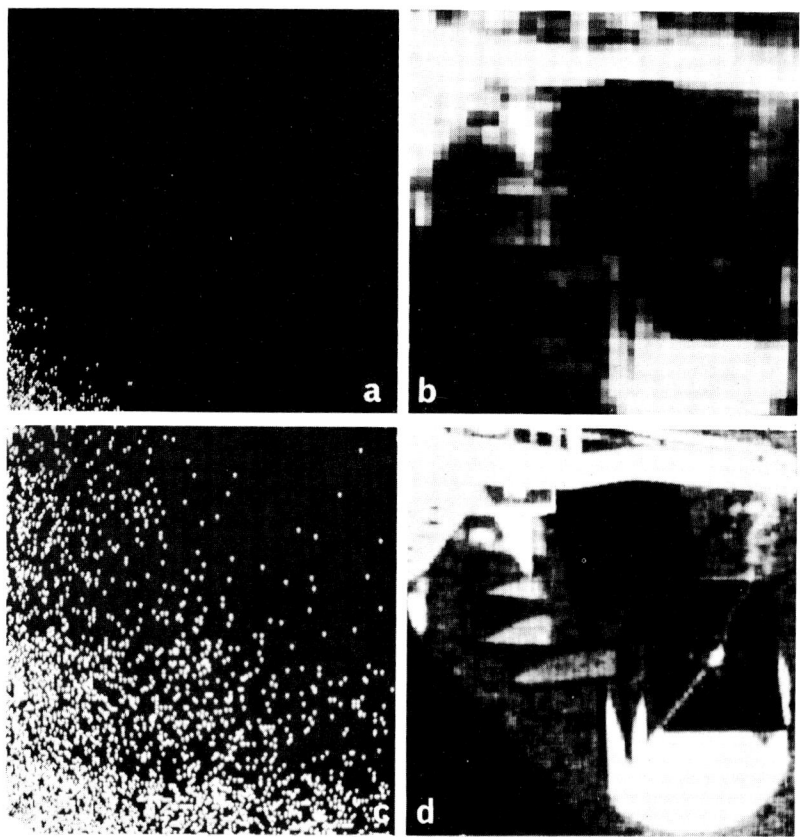

Fig. 17. Hadamard transform thresholded sampling with high and low thresholds, box. (a) Intolerable threshold 500:1 sample reduction. (b) Hadamard transform of thresholded samples. (c) Tolerable threshold 10:1 sample reduction. (d) Hadamard transform of thresholded samples.

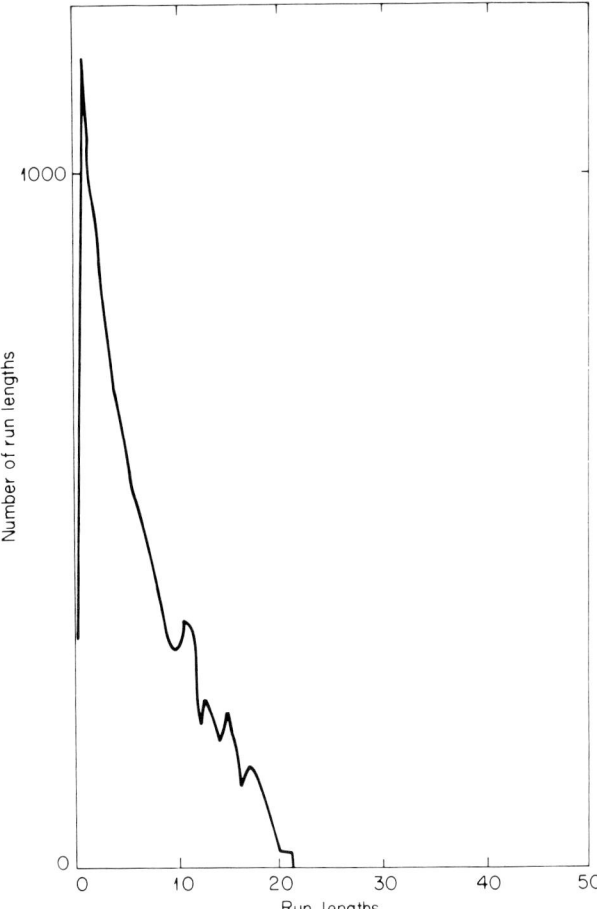

Fig. 18. Run lengths of significant samples for Fourier transform threshold sampling of box at level 200.

170 IMAGE CODING

In order to achieve a bandwidth reduction with this threshold technique of sample deletion it is necessary to code the positions of the significant samples as well as their values. Position coding, of course, adds to the transmission bandwidth. Statistical data has been obtained on the number and location of significant samples in order to determine useful codes and evaluate the amount of bandwidth reduction possible. Figure 18 is a plot of run lengths of significant samples of the Fourier transform of the box for the threshold set at level 200. For this scene the number of run lengths greater than 16 elements is small so that the run lengths can be truncated to 16 without appreciably affecting the distribution. For a sample run length position code with a constant word length of 4 bits, a bandwidth reduction of greater than 4:1 is possible for this scene. A Huffman variable length code would result in a slightly higher bandwidth reduction factor.

7.3 ERROR TOLERANCE

The major advantage of image transform coding other than its potential for bandwidth reduction is the tolerance to channel errors that transform coding affords. The inherent "error averaging" property of transform coding combined with error correction coding of transform samples provides a means of image coding for which channel errors are less deleterious than for conventional spatial coding of an image.

To illustrate the error tolerance feature of transform coding, a binary symmetric channel will be assumed as a model for the channel. In

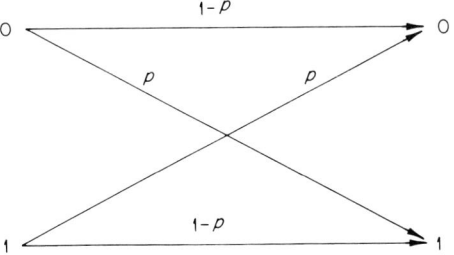

Fig. 19. Model of a binary symmetrical channel.

the binary symmetric channel shown in Fig. 19 the probability of receiving an incorrect symbol is given by p for the transmission of 1's or 0's.

7.3.1 Channel Noise Effects

An intuitive justification for transmitting the frequency rather than the spatial domain of an image is the fact that for many transforms the channel noise introduced in the transform of an image tends to be distributed evenly over the entire reconstructed image. Consequently the noise manifests itself as a low frequency effect in reconstruction. Since the eye is more sensitive to the high frequency "salt and pepper" effect of channel noise in the spatial domain, the same channel noise power in the frequency domain is somewhat less offensive. Figure 20(a) shows a mid-gray scene after having passed through a channel with probability of error of 0.1. Figure 20(b) is the Fourier transform of the output of the same channel whose input was the Fourier transform of the mid-gray scene. Both scenes have the same amount of noise energy but that energy is distributed quite differently. A quantizing and coding method can be developed to take advantage of the inherent high frequency or "salt and pepper" noise immunity that Fourier domain coding offers. As a first step in this direction, a requirement will be made that each quantum level occur equally likely as any other quantum level. This quantization criterion will guarantee that each code word is equally likely to occur and will avoid any unexpected noise biasing, since the binary symmetric channel effects each code bit, and therefore each code word, independently of all others.

Figure 21 contains a series of experimental results for the Fourier transform of the footpad scene using the Gaussian quantization law with the transform domain variance changing as a function of frequency according to the power spectrum of the original scene. The footpad and its quantized Fourier transform are passed through the same binary symmetric channel for two different error probabilities. These pictures are presented to demonstrate a further complication that must be avoided. The frequency induced noise energy is concentrated in low frequency variations which are so large that the high

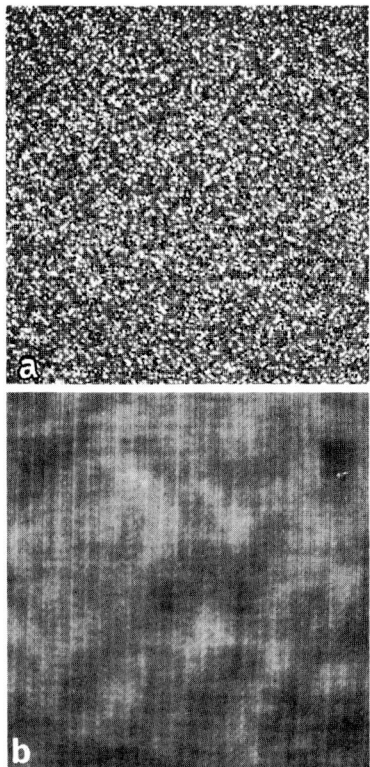

Fig. 20. Binary symmetric channel noise with error rate $p = 10^{-1}$. (a) BSC noise in spatial domain. (b) Fourier transform of BSC in Fourier domain.

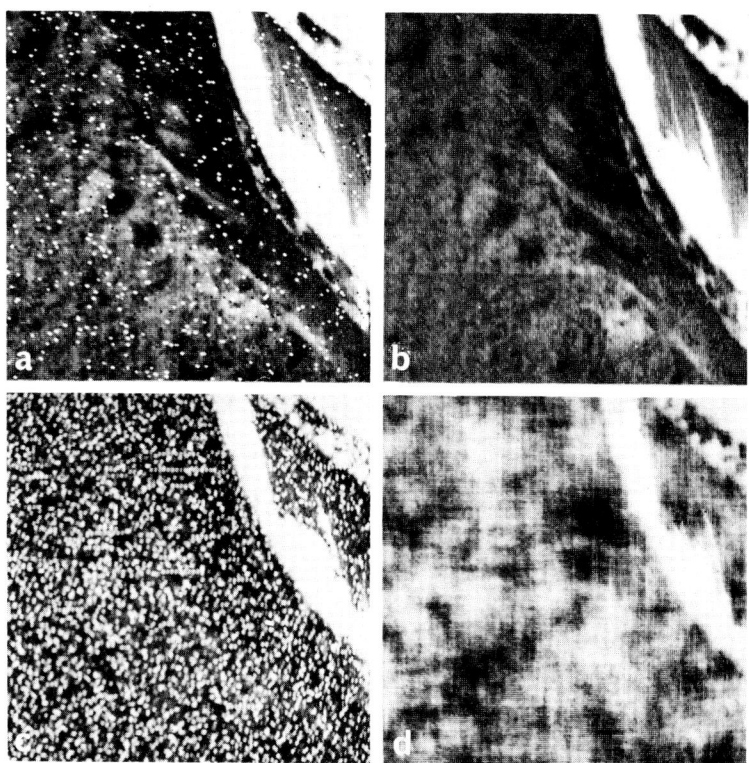

Fig. 21. Binary symmetric channel noise in spatial and Fourier domain transmission. (a) 10^{-3} error rate in the spatial domain. (b) 10^{-3} error rate in the Fourier domain. (c) 10^{-1} error rate in the spatial domain. (d) 10^{-1} error rate in the Fourier domain.

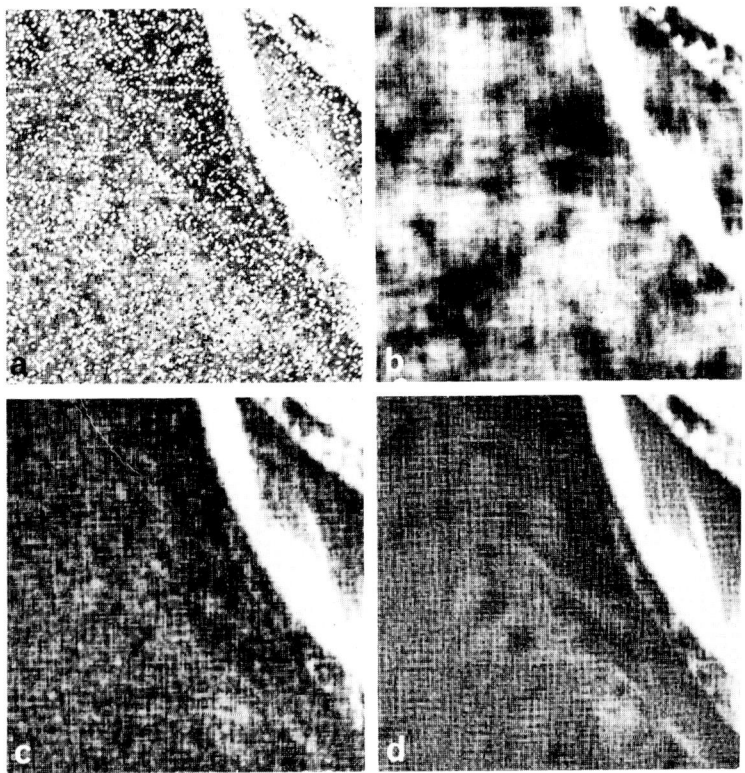

Fig. 22. Effect of low frequency errors. (a) 10^{-1} error rate in the spatial domain. (b) 10^{-1} error rate in the Fourier domain. (c) Reconstruction with the 800 lowest spatial frequencies errorless. (c) Reconstruction with the 6500 lowest spatial frequencies errorless.

frequency information is lost due to normalization in reconstruction. This can be explained by the fact that the absolute, as opposed to the relative value of a bit error is much larger in the regions where the power spectrum is large. In the power spectrum of most images, the larger values occur at the lower frequencies, and thus the lower frequency noise errors have a greater effect on the reconstructed image in the spatial domain. Further demonstration of this effect is afforded by Fig. 22. Figures 22(a) and 22(b) are the footpad noise scenes with error rates of 10^{-1} introduced in the space and frequency domain respectively. Figure 22(c) is the result of the same error rate channel noise in the frequency domain but with 20×40 or 800 of the lowest spatial frequencies transmitted error free. It is evident from Fig. 22(d) that the noise energy is now concentrated in the higher frequencies. Figure 22(d) has the lowest 6500 spatial frequencies transmitted error free.

7.3.2 Error Correction Transform Coding

As a result of the statistical regularity of samples in the frequency domain, a much smaller amount of error correction in this domain will yield a far better noise immunity than the same amount of error correction in the spatial domain. The nature of the quantization law is such that errors in certain positions of the frequency domain are much more bothersome than in other positions due to the large statistical variance of samples at these frequencies. Therefore, it is natural to develop an error correction rule to correct for errors only in these large variance regions. One such rule would be to error code those frequency samples which correspond to positions in the frequency domain where the power spectrum of the covariance function indicates a high probability of large sample values. This technique alone requires an increase in bandwidth to facilitate the error correction. However, it has been found that the small increase in bandwidth in the Fourier domain will result in far better reconstructions than the same increase in the spatial domain.

It is important to emphasize that the coding technique used for the Fourier domain should be tailored to a particular channel capacity. If the channel noise has an error rate less than about 10^{-3}, then it ap-

176 IMAGE CODING

pears that no error correction is necessary as in Fig. 21(b). However, under the circumstances of a high error rate, it often becomes more desirable to transmit as many error corrected samples as possible at the expense of not transmitting the entire frequency plane. Using such a system, corrected, but not necessarily errorless data could be received until normal picture bandwidth has been reached, at which time transmission is terminated. In order to implement such a scheme, an error correcting code must be selected. The code selected will depend on how much of the frequency domain will be omitted due to the increased error correcting capability of the code. The main point of this discussion is to illustrate the variety of coding implementations possible for different channel conditions.

A specific example of the potential of the Fourier coding technique is presented below. A high error rate channel is assumed with rate $p = 4 \times 10^{-2}$. The equal bandwidth criterion is assumed. Consequently, the Fourier coding technique requires exactly the same bandwidth as conventional spatial domain transmission systems. The error correcting code must have at least six information bits. Two such codes which become candidates for implementation are a first-order Reed–Muller code and a Bose–Chaudhuri–Hocquenghem (BCH) code [4, 5, p. 163]. The particular Reed–Muller code of interest is a (32, 6) code in which the minimum distance between code words is 16, and therefore, the code is capable of correcting a total 7 errors. The BCH code is a (31, 6) code and is also capable of correcting errors. The BCH code will be used in the following discussion. Utilizing an error correcting code capable of seven error corrections does not mean that the 6 information bits will be received over the noisy channel error free. Since each code word length has been increased to 31 bits, 8 or more errors per code word cannot be guaranteed to be corrected. The probability of having 8 or more errors in the BCH code is given by the partial sum of the binomial distribution

$$P(8 \text{ or more errors}) = \sum_{i=8}^{31} \binom{31}{i} p^i (1-p)^{31-i} \qquad (54)$$

where p is the binary symmetry channel error rate. This probability is an upper bound for the incorrect reception of a code word since the possibility of correct reception for greater than 7 error still exists

7.3 ERROR TOLERANCE 177

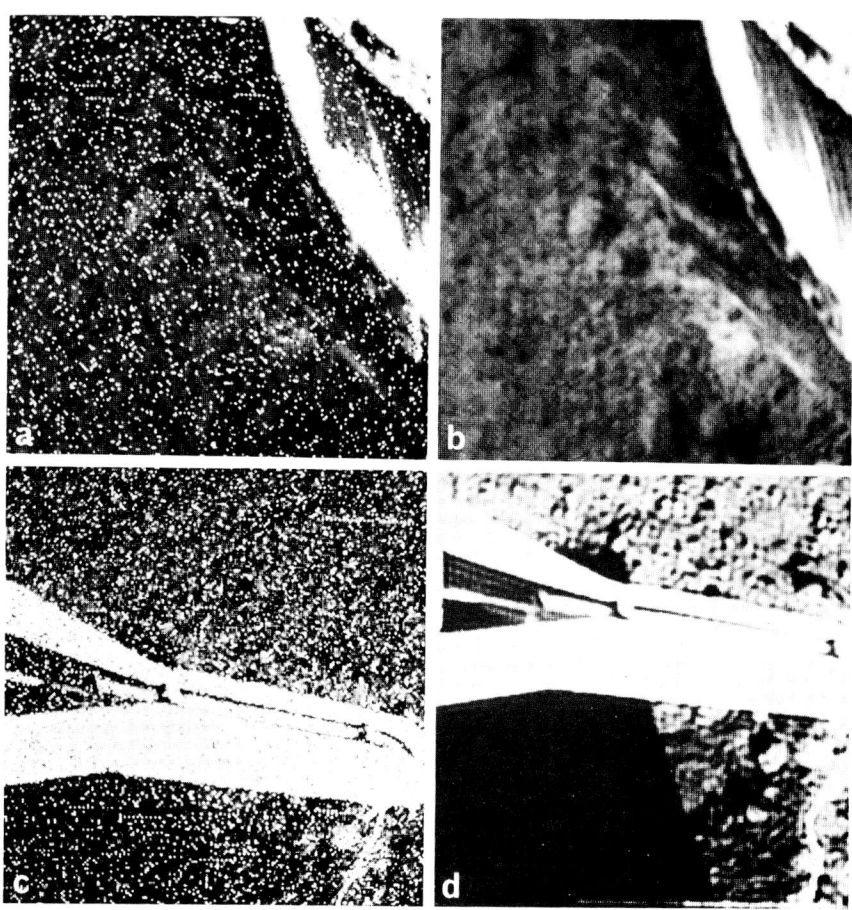

Fig. 23. Equal bandwidth error correction technique. (a) 4×10^{-2} error rate in the spatial domain. (b) Error corrected retransformation. (c) 4×10^{-2} error rate in the spatial domain. (d) Error corrected retransformation.

but is unknown. For the specific channel error rate of 4×10^{-2}, the the error corrected data samples will be received with probability of error no greater than 2.26×10^{-5} [6]. Figure 23 displays the results of this error correcting procedure. Figures 23(a) and 23(c) are two test scenes whose spatial domains are transmitted through the binary symmetric channel with the above error rate. Figures 23(b) and 23(d) are the error correction Fourier domain transmission results for each of the test scenes. While there is a loss of high frequency information in Figs. 23(b) and 23(d) there is a marked improvement over the spatial coding in Figs. 23(a) and 23(c). It is evident that this particular type of coding offers a considerable advantage for very noisy communication channels.

7.4 CONCLUSIONS

This chapter has presented various techniques for using the transform domains of images in a digital communication system. It has been experimentally verified that it is possible to quantize the transform domain of an image in such a way that no image degradation results and the bandwidth (number of coding bits) requirement for the transform domain is the same as for the spatial domain. Bandwidth reduction techniques were then presented demonstrating the possibility of large reduction ratios utilizing the transform information. Both zonal and threshold sampling schemes were presented with experimental results where applicable. The large bandwidth reductions were due, in part, to the compacting-of-image-energy property of both the Fourier and Hadamard transforms. Next the transform domains were investigated as to their potential for noise immunity coding and examples were presented for a binary symmetric channel noise model and fairly large error rates. Again due to the relatively few coefficients containing a large percentage of the total image energy, local error correcting coding results in a global noise immunity upon image retransformation.

REFERENCES

1. H. C. Andrews, "Entropy Considerations in the Frequency Domain," *Proc. IEEE Letters* **46,** No. 1, 113–114 (January 1968).
2. P. F. Panter and W. Dite, "Quantization Distortion in Pulse Count Modulation with Nonuniform Spacing of Levels," *Proc. IRE* **39,** No. 1, 44–48 (Jaunary 1951).
3. P. F. Panter, *Modulation, Noise and Spectral Analysis Applied to Information Transmission.* McGraw-Hill, New York, 1965.
4. I. S. Reed, "A Class of Multiple Error-Correcting Codes and the Decoding Scheme," *Trans. IRE* **PGIT-4,** 38–49 (1954).
5. W. W. Peterson, *Error Correcting Codes.* MIT Press, Cambridge, Massachusetts, 1961.
6. *Tables of the Binomial Probability Distribution.* Dep. Commerce, Nat. Bur. Stand. Appl. Math. Ser . No. 6 (January 1950).

AUTHOR INDEX

Numbers in parentheses are reference numbers and indicate that an author's work is referred to although his name is not cited in the text. Italic numbers refer to the page on which the complete reference appears.

A

Alexits, G., 86(20), *103*
Algazi, V. R., 100(30), *103*
Andrews, C. A., 123(35), *133*
Andrews, H. C., 33(2), 41(6), *54*, 73(6), 76, 81(16, 17), 90, 93(16), 96(17), *102*, *103*, 105(8-12), 106(13-15), 116(9-12), 118(9), 135(1), *179*
Armitage, J. D., 24(18), *29*

B

Barakat, R., 21(12), *29*
Barnes, C. W., 27(33), *30*
Baumert, L., 122, *132*
Bellman, R., 77(14), *103*
Bernfeld, M., 56(3, 5), *70*, *71*
Bershad, N. J., 27(35), *30*
Bochner, S., 127(39), *133*
Boulton, P. I. P., 100(31), *103*
Bremermann, H. J., 98(24), *103*
Brigham, E. O., 105(7), *132*
Brown, B. R., 63(15), *71*
Brown, H. A., 27(34), *30*
Brown, W. M., 57(8), *71*

C

Caspari, K., 90, *103*
Chrestenson, H. E., 81(18), *103*
Cochran, W. T., 73(4), 87(2), *102*, 105(5), *132*
Cook, C. E., 56(3, 5), *70*, *71*
Cooley, J. W., 73(1), 76(1), 87(1), *102*, 105(4, 6), 116(6), *132*
Cooper, G. R., 27, *30*
Croce, P., 59(10), *71*

D

Davenport, W. B., Jr., 27(31), *30*, 56(2), *70*, 62(11), *71*, 75(10), *102*, 123(31), *133*
Davies, J. M., 123(35), *133*
De, M., 63(13), *71*
DeVelis, J. B., 50(8), *54*
Diananda, P. H., 130(42), *133*
di Francia, G. T., 25, *30*
Dite, W., 146(2), *179*
Dramer, H. P., 123(29), *133*

F

Fine, N. J., 122(24, 25), *132*, *133*
Flammer, C., 26(28), *30*

Frieden, R. B., 27(32), *30*

G

Gentleman, W. M., 125(37), *133*
Gnedenko, B. V., 129(40), *133*
Golomb, S. W., 120(18), 122(21), *132*
Good, I. J., 76, 82, *102*
Goodman, J. W., 6(1), 21(1), *28*, 12(7), 18(9), 24(19, 22), *29*, 25(26), 27(26), 36(3), *30*, *54*
Goodman, L. M., 123(34), *133*
Guinn, D. F., 76(13), 90(13), *102*, 123(30), *133*

H

Haar, A., 85, *103*
Hadamard, J., 120(16), *132*
Hall, M., Jr., 122(21), *132*
Harmuth, H. F., 93(23), *103*, 122(22), *132*
Harris, J. L., 22(15), 25(27), 27(27), *29*, *30*
Henderson, K. W., 122(27), *133*
Herrick, R. B., 24(18), *29*
Horowitz, H., 123(36), *133*
Huang, J. J. Y., 123(32), *133*
Huggins, W. H., 123(32), *133*

I

Ingalls, A. L., 41(5), *54*

J

Jacobs, I. M., 56(4), *70*

K

Kane, J., 76, 81(16), 93(16), *103*, 106(13), *132*

Kelly, D. L., 63(12), *71*
Klooster, A., Jr., 57(6), *71*
Kozma, A., 63(12), *71*

L

Landau, H. J., 27(30, 36), *30*
Landgrebe, D. A., 27, *30*
Lewis, P. A., 105(6), 116(6), *132*
Lighthill, M. J., 6, *29*
Loeve, M., 127(38), *133*
Lohmann, A. W., 22(16), 24(18, 23), *29*, 58(9), 63, *71*

M

Marechal, A., 59(10), *71*
Mathews, M. V., 123(29), *133*
McGlamery, B. L., 20(11), *29*
McLaughlin, J. A., 100(29), *103*
Mertz, L., 6(4), *29*
Michelson, A. A., 43(7), *54*
Morgenthaler, G. W., 122(26), *133*
Morrow, R. E., 105(7), *132*
Mueller, P. F., 20(10), 40(4), *29*, *54*

O

O'Neill, E. L., 6(3)
Oppenheim, A. V., 23(17), *29*

P

Palermo, C. J., 123(36), *133*
Palermo, R. V., 123(36), *133*
Paley, R. E., 122, *132*
Panter, P. F., 146(2, 3), *179*
Papoulis, A., 6(2), 11(6), 22(13), 24(21), 27(38), *29*, *30*, 74(7, 8), *102*, 130(43), *133*

Paris, D. P., 24(23), *29*, 58(9), 63(14, 17), 71
Pease, M. C., 73(2), 87(2), *102*
Peterson, W. W., 176(5), *179*
Pollak, H. O., 26, 27(29, 30, 36), *30*, 75(9), *102*
Pratt, W. K., 33(2), 41(6), *54*, 81(16, 17), 93(16), 96(17), *103*, 105(1, 10 12), 106(13), 116(10–12), 118(9), *131*, *132*

R

Rademacher, H., 122(28), *133*
Rao, C. R., 100(26), *103*
Raviv, J., 100(28, 29), *103*
Reed, I. S., 176(4), *179*
Reynolds, G. O., 20(10), 40(4), *29*, *54*
Root, W. L., 27(31), *30*, 56(2), 62(11), *70*, *71*, 75(10), *102*, 123(31), *133*
Rosenfeld, A., 105(2), *131*
Rotz, F. B., 57(6), *71*
Rozanov, Y. A., 130(41), *133*
Ryser, H. J., 120(17), *132*

S

Sakrison, D. J., 100(30), *103*
Schroeder, M. R., 22(14), *29*
Schutheiss, P. M., 123(32), *133*
Schwarz, G. R., 123(35), *133*

Selzer, R. H., 31(1), *54*
Slepian, D., 26, *30*, 75(9), *102*
Streeter, D. N., 100(38), *103*

T

Tukey, J. W., 73(1), 87(1), *102*, 105(4), *132*
Turin, G. L., 56(1), *70*

V

Vander Lugt, A., 16, 17, 24(20), *29*, 57(6, 7), 60, *71*

W

Walsh, J. L., 81(15), *103*, 122(23), *132*
Watanabe, S., 100(27), *103*
Watari, C., 87, *103*
Welch, P. D., 105(6), 116(6), *132*
Werlich, H. W., 63(14), *71*
Whelchel, J. E., 76(13), 90(13), *102*, 123(30), *133*
Wilks, S. S., 100(25), *103*
Williamson, J., 122, *132*
Wozencraft, J. M., 56(4), *70*

Y

Young, T. Y., 123(33), *133*

SUBJECT INDEX

A

Aberration, 19-21
  free, 19
Activity, 100
Airy disk, 42
Algorithm, 32-33, 77
  learning, 98-101
Analytic continuation, 25, 36
Antenna diffraction patterns, 2, 40-50
  supergain, 25
Aperture, 13, 18, 22, 26, 40-50, 53
Apodization, 27
Autocorrelation, 9, 25

B

Bandlimited, 26
Bandwidth reduction, 2-3, 75, 81, 99, 105, 113, 123, 131, 135, 151-170
Bessel functions, 10-11, 22, 42, 74
Binary symmetric channel, 3, 135, 170-171
Bochner's theorem, 127
Boolean, 79-80
Bose-Chandhuri-Hocquenghem code, 176

C

Central limit theorem, 129
Channel capacity, 135
Channel noise, 171-175

Coherent illumination, 2, 12, 16, 18, 26, 28, 60
Coherent transfer function, 7, 18-21, 28
Communication systems, 55-56, 131
Contrast reversal, 36
Convergence, 81
Convolution, 7, 9, 15-16, 22-23, 26, 28, 31, 58, 74
  integral, 7
  theorem, 7

D

Degradation, 20
  aberrations, 20
  contrast, 20, 28
  defocus, 21-22, 28, 36, 53
  motion blur, 20, 22, 28
  turbulence, 20-21
Dictionary sequence, *see* Lexicographic
Diffraction, 2, 12-15, 28, 53, 63
  computer calculated, 40-50, 63
  Fraunhofer, 12, 14-15, 28, 43
  Fresnel, 12, 14-15, 17, 28
  Huygens' Fresnel principle, 12
  Rayleigh-Sommerfeld model, 12, 14, 28
Diffraction limited, 12, 18-20, 25, 36
Digital Fourier holography, 2, 50-53
Digital optical processing, 31-54
Dirac delta function, 6
Dynamic range, 130, 151

SUBJECT INDEX **185**

E

Edges, 58-59, 66, 69
Eigenvectors, 3, 10-12, 26-27, 74-75, 101, 116, 123-124, 162
Entropy, 135-136, 140
Ergodic, 56, 129-130
Error, 27
 channel, 105, 113, 131, 135
 correcting codes, 74, 81, 170, 175-178
 criteria, 138
 Gaussian error function, 140, 163
 mean square, 66, 75, 125
 mean square truncation, 27
 quantization, 136
Error tolerance, 170-178

F

Fast transformations, 76-90
Feature selection, 74, 76, 98, 102
Feature space, 99, 101
Film, 20, 50
Filters
 contrast, 23-24
 Gaussian, 20
 inverse, 20-23, 28, 36, 40, 53
 phase grating, 24
 Schlieren, 24
 Whitening, 58
 Zernike's, 24
 zero-one, 33, 53
Finite dimensional vector spaces, 73
Flying spot scanner, 33
Fourier hologram, 50-53
Fourier optics, 2, 5-19, 28
Fourier transform, 3, 15, 17-18, 22-23, 26, 28, 31-33, 36, 42, 50, 53, 60-61, 75, 87, 92, 96, 98, 101, 113, 116-120, 122, 126, 131, 136-137, 140, 151, 156, 161, 168
 conjugate symmetry, 117-118
 Fourier series, 118-119

G

Gaussian distribution, 136, 150, 162, 163
Gaussian quantizer, 140, 150
Generalized Haar transform, 75, 87
Generalized Walsh transforms, 75, 81-82, 84, 96, 101
Ghost, 23
Gradients, 10-11, 23, 55-56, 59, 64-67

H

Haar transformation, 75, 85-86, 101
Hadamard transform, 3, 75, 80-83, 93, 96, 98, 100-101, 105, 113, 116, 120-123, 130-131, 136-137, 151, 161, 168
Hankel transforms, 10-12, 74
Hard-limiting, 62
Heisenberg's uncertainty principle, 27
Holography, 2, 50-53, 60, 70
Huffman code, 170

I

Image coding, 135-178
Image detection, 55, 64-65
Image enhancement, 2, 5, 19-28, 36-40, 53, 56
Image transforms, 105-131
Imaging systems, 15-19
 coherent, 15-18
 incoherent, 18-19
Impulse response, 7, 18, 56-58, 74, *see also* Point spread function
Incoherent illumination, 2
Inner products, 73
Inverse probability, 55-56

J

Jacobian, 135

K

Karhunen-Loeve expansion, 27, 75, 123-125, 131
Kernel, 114, 116, 131
  Fejer, 127
  separable, 114, 124, 131
  symmetric, 114, 131
Kronecker delta function, 78, 129
Kronecker product, 3, 76-83, 87, 90, 93, 101, 116

L

Lagrange multipliers, 143
Laplacian operator, 23
Lens, 15, 31
  conical, 15
  convex, 15, 17
  cylindrical, 15
Lexicographic, 77, 81, 93
Likelihood ratio, 55-56
Linear operators, 73, 96
Linear systems theory, 2, 6, 10, 17, 23, 28, 31
  two-dimensional, 5-12
Logarithmic quantizer, 140, 150

M

Matrix, 31, 73, 119
  circulant, 31
  core matrix, 78, 87, 90
  covariance, 75
  factorization, 76
  Good matrices, 82-83, 90
  inversion, 74
  orthogonal, 79, 80, 115, 124.
  submatrix, 76
  symmetric, 80, 120
  transpose, 74
  unitary, 115
Maximizing variances, 99-100
Maxwell's equations, 12

N

Neumann factor, 62, 127
Noise immunity, 2-3
Normal distributions, 129

O

Optical data processing, 5-30
  geometrical optics, 12, 22
  physical optics, 12
Optical transfer function, 19-22, 24, 28, 36, 53, 59
Orthogonal, *see* Transformation
Orthonormal, 26, 70, 120

P

Parseval's theorem, 9, 11, 130
Pattern recognition, 64, 74, 98-101
Pattern space, 98
Point spread function, 7, 18, 20
Power spectrum, 56, 58, 66, 126, 136
  amplitude, 63
  phase, 63
Prolate spheroidal wave function, 25-28, 74
Principal component analysis, 76, 100
Prototypes, 99, 100
Pupils, 18, 20, 25-26, 28
  entrance, 18
  exit, 18

Q

Quantization, 3, 106, 135, 136-151
  levels, 142-143, 145, 150
Quantum analysis, 96-98

R

Radar systems, 55
Rademacher functions, 122
Rayleigh diffractions limit, 25
Rayleigh distribution, 136, 162
Redundancy, 76, 82
Reed-Muller code, 176
Resolution, 25, 31, 157
Riemann integration, 127
Ringing, 40

S

Sampling theory, 25-26, 113, 126, 151
Sequency, 93, 122-123, 131
Shift invariant, 79, see also Space invariant
Signal to noise ratio, 55-57
Space invariant, 7
Spatial correlation, 113, 128
Spatial filtering, 31, 33
Spatial frequency, 18, 22, 25-26, 36, 53-54, 62, 116, 119, 157
Spectral analysis, 2-3, 73
  generalized, 76, 78, 90-98, 115
Spectral energy, 33
Statistical analysis, 126-131
Stochastic approximations, 74, 128
Superposition, 6-7
Superresolution, 2, 20, 24-28, 36, 74
Surveyor spacecraft, 40

T

Threshold sampling, 3, 161-170
Transducer, 98
Transform sampling, 151-156
  checkerboard, 153, 156
  random, 153, 156
Transformations, 3
  fast Fourier, 3, 31, 76, 82, 87, 105-106, 123
  fast Hadamard, 3, 76, 82, 106
  Hankel, 10
  Hilbert, 24
  Kronecker products, 3
  orthogonal, 3, 10, 70, 73-102
  symmetric, 18
Two-dimensional matched filters, 2-3, 23, 31, 55-70
  gradient matched filters, 2, 23, 56, 59, 64-66, 70
  image evaluation, 2, 56, 59, 66-69
  Laplacian matched filter, 59

V

Vidicon, 33

W

Walsh, 3, 75, 80-83, 86-87, 92, 93, 96, 98, 100-101, see also Hadamard transform
  functions, 122
Wide sense stationary, 56, 126

Z

Zonal sampling, 3, 157-161